ANTIOXIDANT SUICIDE:
Excessive Antioxidant Intake Linked to Increased Mortality

Prof Randolph Michael Howes MD PhD
Physician, Surgeon
and
Scientist (Molecular Biologist/Biochemist)

Adjunct Assistant Professor of Plastic Surgery, (RET.)
The Johns Hopkins Hospital, Baltimore, MD, USA

Espaldon Professor of Plastic and Reconstructive Surgery,
University of Santo Tomas, Manila, Philippines

Adjunct Professor of Biological Sciences,
Southeastern Louisiana University

Founder, Director and Chairman of the Scientific Advisory Board:
U.S. Medical Scientific Research Foundation, Inc.

By: Dr. Randolph "HealR" Howes

Address for Communication: 27439 Highway 441, Kentwood, Louisiana 70444-8152 USA.
Email: rhowesmd@hughes.net

Notice to Users:

It is understood that medicine is an ever-changing science. As new research and clinical experience broaden our knowledge, changes in treatment and drug therapy are required. The author and the publisher of this work have checked with sources believed to be reliable in their efforts to provide information that is complete and generally in accord with the standards accepted at the time of publication. However, in view of the possibility of human error or changes in the medical sciences, neither the authors nor the publisher nor any other party who has been involved in the preparation of publication of this work warrants that the information contained herein is in every respect accurate or complete, and they disclaim all responsibility to any errors or omissions or for the results obtained from use of the information contained in the work. Readers should confirm the information contained herein with other sources. For example, and in particular, readers are advised to check the product information sheets (or labels) included in the package of each drug they plan to administer to be certain that the information contained in this work is accurate and that changes have not been made in the recommended dose or in the contraindications for administration. This recommendation is of particular importance in connection with new or infrequently used drugs, additives or supplements.
Disclaimers: Please note: only your personal physician or other health professional you consult can best advise you on matters of your health based on your medical history, your family medical history, your medication history, and how information from any of these databases may apply to you. Neither Dr. Howes nor any party involved in creating, producing or delivering this web site shall be liable for any damages arising out of access to or use of this material or web site, or any errors or omissions in the content thereof.

The information given herein is not intended as medical advice. Always consult with your doctor for underlying illness. Before beginning dietary investigation, consult a dietician or a physician with an interest in nutrition. Information is drawn from the scientific literature, web research, and personal enquiry; while all care is taken, information is not warranted as accurate and the author cannot be held liable for any errors and omissions.

Disclaimer: The information provided herein is stated to be truthful and consistent, in that any liability, in terms of inattention or otherwise, by any usage or abuse of any policies, processes, or directions contained within is the solitary and utter responsibility of the recipient reader. under no circumstances will any legal responsibility or blame be held against the publisher or author for any reparation, damages, or monetary loss due to the information herein, either directly or indirectly.

The information herein is offered for informational purposes solely and is universal as so. The presentation of the information is without contract or any type of guarantee assurance.

About the Author
Randolph M Howes MD PhD
Biographical Sketch:

Professor Randolph M. Howes M.D., Ph.D. was born on August 17, 1943 in a small rural hospital in Madisonville, Louisiana. While at the hospital, an accidental third-degree hip burn from a heating pad introduced Dr. Howes to the harsh realities of life.

Raised on a small bucolic strawberry farm in Ponchatoula, Louisiana, Dr. Howes learned ethics, morality, hard work and respect for his fellow man at a young age. Humble beginnings launched his lifetime trek of achievement as a scientist, surgeon, molecular biologist, prolific writer, scholar, visionary, philanthropist, international lecturer, singer, songwriter, business entrepreneur, broadcaster, inventor, author, newspaper columnist, corporate executive and rancher.

He attended St. Joseph's elementary school for eight years, served as an altar boy and sang in the choir. Next, he attended Ponchatoula High School where he finished as President of the Student Council by winning an election over the school's quarterback of the football team. He attributes this hard-fought win to his guitar playing and singing abilities. Dr. Howes began playing self-taught guitar professionally at 13 years of age.

In 1961, Dr. Howes entered Southeastern Louisiana College (now Southeastern Louisiana University, SLU), where he took premed courses, made the Dean's list, made the honors chemistry class, worked 40 hours/week at the Psychology Research Laboratory under Dr. John R. Nichols, played music in his 3-piece combo, named The Three Blind Mice, and was elected as president of the Catholic Youth Organization, the Inter-Fraternity Council and the Junior class. He has been designated a Southeastern "Legend."

He was featured in his college newspaper for his versatility and industriousness and he presented his first scientific paper to the Southwestern Psychological Association on interspecies intelligence, while still in his junior year. He has since been honored as an Outstanding SLU Alumnus, along with Robin Roberts of ABC's Good Morning America. Later SLU articles would refer to him as "a da Vinci in cowboy boots" and the sobriquet stuck.

He has served as an Adjunct Professor of Biological Sciences for many years at SLU. His Southeastern Louisiana University education opened the doors of academia for him and he next matriculated to Tulane School of Medicine in New Orleans, Louisiana.

While working on double doctorate degrees, Dr. Howes worked as a technician on the isolation of thyrotropin releasing factor with Nobel Laureate, Dr. Andrew V. Schally, studied under Dr. Richard Steele, whose mentor was Nobel Laureate, Dr. Albert Szent Gyorgii, met Nobel Laureates, George Wald and Dr. Linus Pauling, who felt that Dr. Howes could help bridge the gap between physicians and scientists, served as president of the Biochemistry graduate students, graduated in the top 10, received the 1971 Pathology Association Award, was elected to Sigma Xi honor fraternity and was the first in the history of Tulane School of Medicine to receive double doctorate degrees in medicine and biochemistry simultaneously.

He was the first to be designated by the late Dr. Theodore Drapanas as a trained "surgical scientist" at Tulane Medical School.

He matched with his first choice at the prestigious Johns Hopkins Hospital for his surgical internship and residency training. He chose it over other top-notch programs because Dr. George Zuidema, Chief and Blalock Professor of Surgery, gave him permission to conduct research studies concurrent with his surgical training in the highly sought-after William Stewart Halstead surgical training program.

Even during his internship year, he was permitted lab space by Dr. John Cameron, past president of the American Surgical Association, and he secured his own grants, trained his own lab technicians, slept on the lab floor and later wrote many papers on surgical and oxygen free radical subjects during his residency training.

At the end of his second year of training, he was selected (as only the second resident to do so) to enter a double training surgical program in general and plastic surgery, to be completed in a six-year interval.

He played music and sang for many of the surgery resident's functions and broke an ankle in a resident's football game and sustained significant trauma from a motorcycle accident, neither of which caused him to miss a single minute of work.

He was the first to complete board eligibility in both general and plastic surgery at The Johns Hopkins Hospital, while doing basic research on oxygen metabolism, all in a six-year period.

He had the opportunity to work with the pioneer of mitochondrial biochemical function, Dr. Albert L. Lehninger, and rubbed elbows with many of the greats of science, surgery and medicine. He trained with Dr. Edward Luce and Dr. James Wells, both of whom have served as president of the American Society of Plastic and Reconstructive Surgeons. He trained under Dr. John E. Hoopes, past president of the American Association of Plastic Surgeons.

He received many grants, honors and awards from 1971-1977 during his years at Johns Hopkins, which are detailed in his full curriculum vitae. He was Dr. Paul Manson's chief resident, who retired as Chief of Plastic Surgery at Johns Hopkins Hospital in 2014. Also, he was granted a position as Adjunct Assistant Professor of Plastic Surgery at Johns Hopkins Hospital, which was another first at Johns Hopkins.

His musical interests have carried him to perform at the New Orleans World's Fair, on many televised shows, appearing with numerous country superstars and ultimately to center stage at the famed Grand Ole Opry Gospel Hour in Nashville, Tennessee. He has composed over 500 songs and his original "Fantasies of You" recording went to the # 1 chart spot on Nashville's Panel Report for nationwide independent air play. His song, "What is Christmas Without You" went to #2 on Tracker Magazine of Nashville.

He was honored by the Country Music Associations of America with a Lifetime Achievement Award, inducted into the Tracker Hall of Fame, received the King Eagle Humanitarian Award for "Your Devotion To the Betterment of Mankind", received the 1999 Golden Music Award, Lifetime Achievement Award for Songwriter/Artist/Humanitarian and many other such honors.

In 1994, he received Dr. Norman Vincent Peale's America's Awards honoring Unsung Heroes, known as "The Nobel Prize for Goodness," and in 1995, he was awarded an Honorary Doctorate of Humanities Degree by SLU. That same year he was sworn in by Rudolph W. Giuliani, Mayor of New York City, as the Community Mayor for the State of Louisiana, International Council of Mayors and was an awardee, along with the late Dr. Stephen Ambrose, for the George Washington Honor Medal.

Even though he had been told that he could not go directly into solo practice, he boldly returned to New Orleans in 1977 and opened his private practice at the Institute of Cosmetic Plastic Surgery, which became a bona fide success story.

He has served as president of the Metropolitan Cosmetic Surgery Society and the Louisiana Cosmetic Surgery Society and has served the American Academy of Cosmetic Surgeons in numerous national offices and in many capacities, including national First Vice President of the American Board of Cosmetic Surgery and chairman of the Board of Advisors for the American Society of Liposuction Surgery.

He was awarded a patent certificate for inventing the triple lumen venous catheter in 1977, licensed it to Arrow International, Inc. in 1981, successfully defended it is a multimillion dollar six-year patent infringement suit and watched it become recognized as the number one venous catheter in the world. His multilumen catheter has been credited with helping save the lives of over 20 million critically ill patients worldwide and the name of Howes is well-known in over 100 countries.

He has had the opportunity to work closely with many of the great minds of medical science. His biochemistry mentor at Tulane was Dr. Richard Steele, whose mentor was Nobel Laureate, Dr. Albert Szent Gyorgii. While at Tulane, he worked as a technician with Nobel Laureate, Dr. Andrew V. Schally, co-isolator of thyrotropin releasing factor.

At Tulane, he had the privilege of meeting and speaking with Nobel Laureate, Dr. Linus Pauling, and Dr. Wald, who was interested in the fact that he was the first in the history of Tulane to complete a Doctor of Medicine and a doctorate in biochemistry simultaneously, which they felt could help bridge the chiasm between physicians and scientists.

While completing his general surgery and plastic surgery training at the Johns Hopkins Hospital, he had the opportunity to work with the late Dr. Albert L. Lehninger, who contributed so much to the study of the mitochondrion and he was able to conduct research on oxygen free radicals throughout his general and plastic surgery residencies.

In the 1980s he became friends with Nobel laureate, Fritz Lipmann, through their work with the Institute of All Nations and the Louisiana University of Medical Science. In later years, he had many fascinating conversations with Nobel Laureate, Dr. Fritz Lipmann, discoverer of co-enzyme-A and its importance in intermediary metabolism.

While at a meeting at Johns Hopkins, he presented an original poem entitled, "Nitric Oxide: Radical Thoughts in vein," to Nobel Laureate, Dr. Louis Ignarro. He feels that exposure to these great minds has been influential in helping him envision new landmark paradigms in the formulation of his innovative research.

He performed pro bono surgery, since 1982 throughout the Philippines, was honored by the Philippine Ministry of Health in 1985 and since 2004, he holds the Espaldon Professorial Chair in Plastic and Reconstructive Surgery at the University of Santo Tomas in Manila.

He was the recipient of the Humanitarian of the Year Award from the Community Mayors of New York, New Jersey and Connecticut in 1996. His philanthropic and humanitarian efforts have been acknowledged by Presidents Ronald Reagan and George H. Bush and he has received a letter of appreciation from past USA President, George W. Bush.

He retired from his private practice to pursue his dream of contributing to a better understanding of oxygen biochemistry and of conducting an arduous in-depth review of the world's scientific literature on oxygen metabolism.

Also, in 2004, in an unprecedented move, The Johns Hopkins Hospital gave him an appointment as an Adjunct Assistant Professor of Plastic Surgery.

His research has proven that the free radical theory is unfounded and that electronically modified oxygen derivatives (EMODs) are of low toxicity and are essential for energy production, for pathogen protection, as secondary cell messengers and as tumoricidal agents.

His Unified Theory states that EMOD insufficiency levels "allow" for the manifestation of diseases, including neoplasia and is a contributing factor in the aging phenomenon. He also postulates that an EMOD insufficiency is the basis for coexistence of diseases.

Dr. Howes, who is both an experimentalist and a theoretician, is an international lecturer on plastic surgery and a world-renowned expert on the biochemistry of oxygen free radicals. His passionate goal is to have cures at the bedside, based on his innovative theories involving electronically modified oxygen derivatives, within his lifetime.

Additionally, even though he does not like talking about it, he is a self-made multi-millionaire.

He is currently doing extensive research on cures for cancer and heart disease and development of revolutionary treatment modalities. He has written 24 books over the past 6 years on the subject of oxygen metabolism, as it relates to protection from cancer, heart disease, diabetes, malaria, HIV/AIDS, Alzheimer's disease, aging and arthritis. He has written many scientific and medical papers and has lectured nationally and internationally.

His research has shown that currently popular antioxidant vitamins, such as vitamins A & E, can be harmful and that oxygen free radicals protect us from bacterial, fungal and viral infections and they help to control cancer growth. He has developed an effective, inexpensive singlet oxygen generating system, from orthomolecular agents, for the treatment of cancer and heart disease.

He has also formulated a proprietary blend of anti-cancer agents to fight and protect against cancer. He is driven by a duty to serve mankind.

There are over 12,000 pages in his magnum opus and he has written over 500 medical editorials in newspapers across Louisiana.

Companion Papers:

Citation: R. Howes: Mythology of Antioxidant Vitamins? *The Journal of Evidence-Based Alternative and Complimentary Medicine.* April, 2011. 16(2): 149-189.

Citation: R. Howes: Cancer Therapy: A Review with Scientific Validation for the Role of Electronically Modified Oxygen Derivatives in Oncologic Treatment Modalities. *The Internet Journal of Alternative Medicine.* 2010 Volume 8 Number 1.

Citation: R. Howes: Hydrogen Peroxide: A review of a scientifically verifiable omnipresent ubiquitous essentiality of obligate, aerobic, carbon-based life forms. *The Internet Journal of Plastic Surgery.* 2010 Volume 7 Number 1.

Howes M.D., PhD., R. (2009). Dangers of Antioxidants in Cancer Patients: A Review. *PHILICA.COM Article number 153.* Published 7th February, 2009. (20 pages)

Howes M.D., PhD., R. (2008). Aging and anti-aging claims: a review on antioxidant vitamins A, C & E. *PHILICA.COM Article number 116.* Published on 12th January, 2008. (16 pages)

Howes M.D., PhD., R. (2007). Sleep: An original "radical" proposal. *PHILICA.COM Observation number 42*. Published on 5th October, 2007. (1 page)

Howes M.D., PhD., R. (2007). Antioxidant Vitamins A, C & E; Death in Small Doses and Legal Liability? *PHILICA.COM Article number 89*. Published on 5th April, 2007. (23 pages)

Howes M.D., PhD., R. (2007). Cancer, Apoptosis and Reactive Oxygen Species: A New Paradigm. *PHILICA.COM Article number 86*. Published on 26th February, 2007. (11 pages)

Howes M.D., PhD., R. (2007). Antioxidant Vitamins A, C and E: Assessing Potential for Harm. *PHILICA.COM Article number 83*. Published on 15th February, 2007. (14 pages)

Howes M.D., PhD., R. (2007). The Consequent Downfall of the Free Radical Theory. *PHILICA.COM Article number 75*. Published on 22nd January, 2007. (9 pages)

Howes, R.M.: "The Free Radical Fantasy," The Annals of New York Academy of Sciences, 2006, Vol. 1067, pp. 22-26.

Available at www.philica.com
www.medi.philica.com
www.iwillfindthecure.org

Companion Books of Prof. R. M. Howes, MD, PhD:

Books online: All books available at www.amazon.com; www.barnesandnobles.com; www.booksamillion.com.

- **Death In Small Doses? Book One:** Antioxidant Vitamins A, C & E in the 21st Century *A Health Impact Statement for The Layman.* **Prof. R.M. Howes MD, PhD.** Trafford Publishing, © **2010**

- **Death in Small Doses? Book Two:** Antioxidant Vitamins A, C & E in the 21st Century *A Health Impact Statement for The Medical Scientist.* **Prof. R.M. Howes MD, PhD.** Trafford Publishing, © **2010**

- **Coffee Table Musings of The DaVinci in Cowboy Boots. Prof. R.M. Howes MD, PhD. Trafford Publishing, © 2010**

- **Antioxidant Overkill, Prof. R.M. Howes MD, PhD.** CreateSpace and Free Radical Publishing, © **2011**

- **Dangers of Excessive Antioxidants in Cancer Patients, Prof. R.M. Howes MD, PhD.** CreateSpace and Free Radical Publishing, © **2011**

- **Heart Disease and Antioxidant Failures, Prof. R.M. Howes MD, PhD.** CreateSpace and Free Radical Publishing, © **2011**

- **Antioxidant Failures and Dangers, Prof. R.M. Howes MD, PhD.** CreateSpace and Free Radical Publishing, © **2011**

- **Anti-Aging Anti-Oxidant Scams, Prof. R.M. Howes MD, PhD.** CreateSpace and Free Radical Publishing, © **2011**

- **Sports, Athletes, Exercise Facts and Antioxidant Myths, Prof. R.M. Howes MD, PhD.** CreateSpace and Free Radical Publishing, © **2011**

- **Alzheimer's Disease: Forget Antioxidants and Supplements, Prof. R.M. Howes MD, PhD.** CreateSpace and Free Radical Publishing, © **2012**

- **Sex, Performance, Reproduction, Naked Radicals and Antioxidants, Prof. R.M. Howes MD, PhD.** CreateSpace and Free Radical Publishing, © **2012**

- **Antioxidants Linked to Deadly Unintended Consequences, Prof. R.M. Howes MD, PhD.** CreateSpace and Free Radical Publishing, © **2013**

- **U.T.O.P.I.A.: Unified Theory of Oxygen Participation In Aerobiosis, Prof. R.M. Howes MD, PhD.** CreateSpace and Free Radical Publishing, © **2014, revised**

- **Hydrogen Peroxide: A Health, Homeostatic and Protective Essentiality, Prof. R.M. Howes MD, PhD.** CreateSpace and Free Radical Publishing, © **2014**

- **Reactive Oxygen Species vs. Antioxidants: The Oxypocalypse or The War That Never Was, Prof. R.M. Howes MD, PhD.** CreateSpace and Free Radical Publishing, © **2014**

- **Diabetes and Oxygen Free Radical Sophistry, Prof. R.M. Howes MD, PhD.** CreateSpace and Free Radical Publishing, © **2014, revised**

- **FISH OIL (Omega3 fatty acids): Facts, Fantasies & Failures. Prof. R.M. Howes MD, PhD.** CreateSpace and Free Radical Publishing, © **2014**

-**Vitamin D: Benefits & False claims. Prof. R.M. Howes MD, PhD.** CreateSpace and Free Radical Publishing, © **2014**

- Chocolate & Red Wine Antioxidants (Polyphenols, Flavonoids & Resveratrol): Facts vs. Falsehoods. **Prof. R.M. Howes MD, PhD.** CreateSpace and Free Radical Publishing, © **2015**

- Blueberry, Tomato & CoQ10 Antioxidants (Anthocyanin, Lycopene & Ubiquinone): Claims vs. Facts. **Prof. R.M. Howes MD, PhD.** CreateSpace and Free Radical Publishing, © **2015**

- Exercise and Reactive Oxygen Species. Likely the only health miracle out there. **Prof. R.M. Howes MD, PhD.** CreateSpace and Free Radical Publishing, © **2015**

- Cancer and Longevity Answers: Naked mole rats, Exercise & EMODs (ROS). **Prof. R.M. Howes MD, PhD.** CreateSpace and Free Radical Publishing, © **2015**

- Hyperbaric Oxygen, Hypoxia, Hyperoxia & EMODS (ROS): Separating Fact from Factitious. **Prof. R.M. Howes MD, PhD.** CreateSpace and Free Radical Publishing, © **2015**

- Pomegranate, Broccoli (sulforaphane) & Turmeric (Curcumin): From the food cart to the medicine cart. **Prof. R.M. Howes MD, PhD.** CreateSpace and Free Radical Publishing, © **2015**

- Cancer Killing, Suppression and Protection: The Howes Answer To Cancer. **Prof. R.M. Howes MD, PhD.** CreateSpace and Free Radical Publishing, © **2016**

Other Books
SCIENCE FICTION:
<u>Slightly Green</u>: **Prof. R.M. Howes MD, PhD.** CreateSpace and Free Radical Publishing**,** © **2011.** (192 pages)

<u>Catching Cancer</u>: **Prof. R.M. Howes MD, PhD.** CreateSpace and Free Radical Publishing**,** © **2012.** (131 pages)

<u>LSU Spirit, in the Mounds</u>: **Prof. R.M. Howes MD, PhD.** CreateSpace and Free Radical Publishing**,** © **2012.** (142 pages)

<u>Pangaea 6</u>: **Prof. R.M. Howes MD, PhD.** CreateSpace and Free Radical Publishing**,** © **2012.** (165 pages)

<u>That Proof of God Thingy</u>: **Prof. R.M. Howes MD, PhD.** CreateSpace and Free Radical Publishing**,** © **2012.** (206 pages)

The Nth Degree: Prof. R.M. Howes MD, PhD. CreateSpace and Free Radical Publishing, © **2012.** (221 pages)

The Saints' Curse and Ghosts on The Levee: Prof. R.M. Howes MD, PhD. CreateSpace and Free Radical Publishing, © **2012.** (154 pages)

Tulane Voodoo: Prof. R.M. Howes MD, PhD. CreateSpace and Free Radical Publishing, © **2012.** (133 pages)

Ultiminium: Prof. R.M. Howes MD, PhD. CreateSpace and Free Radical Publishing, © **2012.** (152 pages)

MOTIVATIONAL:
The Fire Eaters: Molding Your Own Destiny More Easily. Prof. R.M. Howes MD, PhD. Carnivore Press. 1982. Reissued 2014

Books online: All books available at **www.amazon.com** ; **www.barnesandnobles.com** and **www.booksamillion.com**

OTHER BOOKS
PUBLISHED:

Uplift, The Answer Book to your plastic and cosmetic surgery questions, Carnivore Press, © 1986
The Pundit Speaks, An Anthology of Neoclassical Poetic Philosophy, Carnivore Press, © 1990
The Pundit Speaks, Volume II, An Anthology of Neoclassical Poetic Philosophy, Free Radical Press, © 1994
The Pundit Speaks, Volume III, An Anthology of Neoclassical Poetic Philosophy, Free Radical Press, © 1996
The Fable of the Chocolate Covered Strawberry Coloring Book, Free Radical Press, © 2001
The Pundit Speaks, Volume IV, An Anthology of Neoclassical Poetic Philosophy, Free Radical Press, © 2003
The Pundit Speaks, Volume V, An Anthology of Neoclassical Poetic Philosophy, Trafford Publishing, © 2009

DEDICATION

To those who are insanely driven to take on a lifelong quest dedicated to the undaunted task of making the impossible dream a scientific reality.
R.M. Howes, M.D., Ph.D.
5-21-18

The unanswerable mysteries of being alive
are surpassed only by the
incomprehensible mysteries of being dead.
R.M. Howes, M.D., Ph.D.
10-30-11

In the hands of the uninformed,
antioxidant supplements are agents of self-destruction.
R.M. Howes, M.D., Ph.D.
2-26-18

In the hands of supplement marketeers,
antioxidant supplements are agents of mass-destruction.
R.M. Howes, M.D., Ph.D.
3-18-18

It's time to say last rites over Harman's free radical theory.
R.M. Howes, M.D., Ph.D.
2-26-18

Eye Floaters in your seventies:
Let's put it this way, "You are never alone."
R.M. Howes, M.D., Ph.D.
4-23-18

I freely admit to dissecting and destroying
Harman's free radi-crap theory.
R.M. Howes, M.D., Ph.D.
4-26-18

THE FABLE OF
"CLOTHES are KILLERS"
(a parody)

Once upon a half a century ago, investigators studied a group of cancer patients and found that all had worn clothing. It was impossible to deny this obvious and direct association and they questioned why others hadn't made this apparent connection before. It had been known for many years that articles of clothing could be destructive and harmful by getting caught in machinery, producing dismemberment, holding hot liquids and/or strangulating blood supply to various body parts. Scientists set out and verified in the laboratory that clothes could strangle, maim, mutilate, dismember and harm people. Thus, under in vitro laboratory conditions, clothes could destroy the underlying biochemicals of life, i.e., proteins, lipids, and even DNA culminating in the death of the individual. The potential of this horrific scene had been skillfully scrawled in crayon on the nonsense/nonscience canvas of pseudoscience.

Medical researchers obtained grants to study the extent of the clothing menace. They studied a group of patients with atherosclerosis and they had a similar finding of a direct relationship between the wearing of clothes and the occurrence of vascular plaques. Diabetes showed the same thing as did arthritis, Alzheimer's, obesity and cataracts. In fact, all of these diseased patients had worn clothes. The more sick patients they studied, the more they found that all had worn clothing. This was a revelation akin to the discovery of the atom or the circulation of blood. Additional studies verified a straight-line relationship between over 100 pathophysiologies and the wearing of clothing. Study after study demonstrated the same deleterious effect of clothes. The conclusions were obvious, i.e., clothes are damaging, toxic, pernicious, deleterious and harmful, even though they may by needed for survival at times. This conclusion was becoming an established and well accepted fact.

However, some scientists were puzzled by the fact that not all of those who wore clothes actually developed any of these diseases. Consequently, this unsettling observation was immediately dubbed a "paradox" and was shrugged off and swept under the scientific rug of academic cover up. After all, the data was so strong for a direct relationship between the donning of clothes and the occurrence of a vast array of diseases that it had to be a scientifically accurate observation. They had discovered causality of disease.

These deep thinkers found that if they further segmentalized the diseased patients into categories according to the type of fabric, predominant color of their ensemble, mixing of colors, exposure of the individual and the clothing to sunlight or UV, associated smoking, diesel fumes, sex, race, creed or fondness for pets and Twinkies, they could on occasion find statistically significant subgroups for specific diseases. By running the data through the statistical pseudo-scientific meat grinder, they could produce a confusing, virtual sausage of association, which no one would question. This was especially true if they kept the groups small and non-randomized. This mathematically proven revelation inspired the **"clothes cause disease theory."** Thence arose the powerful movement of the very vocal "clothes phobes."

For the next fifty years, the "clothes phobes" continued to dismiss the fact that many bearers of clothing did not become ill at all and that they, in fact, lived normal, healthy long lives. They repeatedly made the point that a paradox here or a paradox there does not dismiss or falsify the fact that all of the studied sick people had, in fact, been wearing clothes on a regular basis and that association was undeniable. Paper after paper seemingly confirmed their apparent assertion, built peer momentum and overrode common sense observations.

In fact, it had become increasingly clear that the longer the people had worn clothes, the more likely it was that they were going to develop illness and to age. Further, it was reasoned that the mere condition of wearing clothes must be producing accumulating stressful and disease causing conditions. It became obvious to them that the random summation of the clothes-wearing-condition was responsible for the onset of diseases and it evidently promoted aging. Thus, was born the **"clothing stress and aging theory."** In fact, one noted West coast biochemist was quoted as saying, "The things which protect our bodies the most, i.e., clothes, are the very things that are killing us." Other respected chemists published **"The Clothes Paradox."**

The clothes phobes widely and emphatically disseminated their message to the extent that anyone doubting their claims of clothing toxicity would be snubbed by the clothes phobe club and labeled a crack pot. Forthrightly, they were boldly quoted as saying, "Even Forest Gump could see that all sick people had worn clothes and that the longer any of them wore clothes, the older they became." It was as clear as the nose on your face or the clothes on your back.

It had been pointed out that some infants were born ill and the clothes phobes explained this anomaly by the fact that the sperm donor was probably a wearer of clothes and such was likely the case for the mother. The unspecified clothing toxic factor was likely passed on to the fetus. Shockingly, it had also been pointed out by some renegades that there were occasions whereby clothes were beneficial and good. Conflicting observations and conclusions were surfacing. In fact, one study found that the wearing of clothes in polar climates could be lifesaving and thus, the confusion grew.

The clothes phobe scientists reasoned that if clothing was the culprit, then illness, disease and aging could be prevented, stopped and/or reversed by not wearing clothes (nudity or an "anti-clothes condition"). Thus, they had millions remove their clothes, followed their clinical course but they unexpectedly observed many "disappointing results." It appeared that there were just as many sick nude people as there were in people who were wearing clothes. In fact, removal of clothing, which could result in parasite attacks, malaria, yellow fever, frost bite, freezing, extreme sunburn, animal bites and overall trauma, could result in an overall increase in disease and mortality.

These observations shook the **"clothes cause disease and aging theory"** to its core. All of the clothes phobe grants were put in jeopardy. However, the clothes phobes struck back with full venomous force. With the injudicious use of statistical smoke and mathematical mirrors, they could still show that clothes and disease and aging were directly related. They repeated the fact that in the lab, items of clothing could readily be made to produce or cause harm to the body and if this was the case, "What could be clearer?" The clothes must go. "Loathe the clothes" was their mantra.

They hastily formed the Committee for Dreaming Up Excuses to refute the ineffectiveness of nudity in prevention or reversal of disease and aging. The Excuse Committee published prolifically, and clever marketers rushed in to sell "virtual see-through clothes" with the implication that this could supplement the nude state. Yet, they were still getting disappointing results. The Excuse Committee produced a litany of pretexts and alibis to negate the failed studies, but it wasn't enough. Clear thinking and common sense prevailed and ushered in the rise and fall of the **"clothes cause disease and aging theory."**

Respectfully (with tongue in cheek) submitted,

Randolph M. Howes, M.D., Ph.D.

p.s. I could have sworn that I had heard something like this before concerning oxygen, but no one could have been that stupid.

News Flash: "Contrary to world orthodoxy, breathing of oxygen has been linked to extremely high mortality rates amongst people with lungs. In fact, all people who breathed oxygen in the 19th century are dead, which is statistical linear proof of its lethality." RMH 10/14/07

**May your life be trouble free
and overflowing with bliss.**
R. M. Howes, M.D., Ph.D.
1-20-17

**Can you recall a revolutionary breakthrough
in our war against cancer?
… neither can I!
Mankind must adopt
prooxidant cancer therapies
for truly innovative advances.**
R.M. Howes, M.D., Ph.D.
3-6-16

**Life is designed such that
human perfection is
never-ever obtainable.**
R. M. Howes, M.D., Ph.D.
1-19-07

Hiding in plain sight
is the glorious fact that oxygen
protects and sustains us.
As happens all too often, we are blinded
by the obvious.
R. M. Howes, M.D., Ph.D.
1-19-07

If the free radical theory had been a horse,
it would now be glue.
R. M. Howes, M.D., Ph.D.
1-13-07

Knowing the truth beats living in blissful ignorance every time.
R. M. Howes, M.D., Ph.D.
4-10-07

Other than the fact that it has been
discredited, refuted, invalidated, disproved, rebutted and negated,
deluded experts believe that the free radical theory
seemingly holds great promise
in the treatment of over 200 diseases
and for the reversal of aging.
Puhleeze….get real, you clowns.
Stop the tomfoolery.
R. M. Howes, M.D., Ph.D.
7-03-07

Antioxidants are free radicals
in sheep's clothing.
These free radical, disguised precursors sneak around
pretending to be unlike the oxidant-wolf
but they equally enjoy feasting on a bloody electron exchange.
Peel back their sheep skin
and there stands a radical.
It is what sustains the nature
and the essence of life.
R. M. Howes, M.D., Ph.D.
11-6-07

Radicals and antioxidants are flip sides
of the same chemical redox coin.
R. M. Howes, M.D., Ph.D.
11-6-07

Test tube biochemical experiments utilize a dead gumbo;
whereas, the breathing human body
is a miraculous, functioning complex
of compartmentalization,
with instantaneous inter- and intra-cellular cross talk,
demonstrating the amazing ability of self-control.
No wonder we cannot generalize the result of one to the other.
Scientists attempt to compare two different beasts:
one, living and the other, dead
and express dismay when their results do not extrapolate.
Can the professed brilliant really
be that dumb?
So,…….. go figure.
R. M. Howes, M.D., Ph.D.
11-6-07

Myths frequently smother facts and
mendacity strangles veracity.
It requires incredible persistence and courage
to expose and teach truth.
In the areas of dietary supplementation and medicine,
data undergoes daily manipulation
to promote "sales based on pseudoscience"
and "profit based on lies."
R. M. Howes, M.D., Ph.D.
11-22-07

Aerobic organisms pose as biochemical Petri dishes
for the study of oxygen metabolism.
We must swirl the dish contents
and try to read the "radical" tea leaves for the future,
….and for the betterment of mankind.
R. M. Howes, M.D., Ph.D.
6-10-07

Customarily, we do not die all at once.
We die in pieces, little by little.
First, the blood vessels of the heart and the brain go.
Concurrently, the lungs lose functionality and
oxygen delivery throughout the body declines,
as we slowly suffocate in the open air.
Muscle mass dwindles, whilst the high school hunk morphs
into a couch-fat-blob.
Vision squints, as silence settles in the ears.
All the while, the plaques thicken and
the cancerous crab crawls and breeds.
Hail, hail, the death gang is all here
to sing the final verse of life.
Peace be with you,
"a little piece at a time."
R. M. Howes, M.D., Ph.D.
1-3-08

"The Oxypocalypse"
Dr. Harman,
"How could so many
be so wrong
for so long?"
R. M. Howes, M.D., Ph.D.
6-6-06

As a species, we suffer mass amnesia
of all of our previous lives.
How quickly we forget.
R. M. Howes, M.D., Ph.D.
1-3-08

Table of Contents:

Disclaimers
About the Author
Companion Papers
Companion Books of Prof. R.M. Howes, MD, PhD
Dedication
THE FABLE OF "CLOTHES are KILLERS"

Prologue

My Story in a Nut Sack

Fundamental Value of Creative Ideas

Antioxidant Skullduggery

Personal Introductory Comments

My Opinion on EMODs and Antioxidant Vitamins
The Howes Doctrine
Insights Gained through my Research

Section One: Mortality and Suicide

Introduction
What are Free Radicals?
What are Antioxidants?
Introductory Discussions on Suicide and Mortality
A List of Major Reactive Oxygen Species and Antioxidant Systems in Living Organisms
Organizations Not Recommending Antioxidant Vitamins
HARMFUL EFFECTS OF ANTIOXIDANTS by Category
MORTALITY (number of deaths) (25 study reports)
CANCER (72 study reports)
HEART DISEASE (28 study reports)
STROKE (6 study reports)
27 Types of Human Cancer Cell Types and 10 Murine Cancers that Can Be Killed by
        EMODs and Protected/Shielded by Antioxidants
SPECIFIC HUMAN CANCER CELL TYPES SHIELDED BY ANTIOXIDANTS

SPECIFIC MURINE (mouse and rat family) CANCER and CELL LINES SHIELDED (52 instances) BY ANTIOXIDANTS
Summary: 34 Human and 10 Murine Cancer Cell Lines Killed by EMOD-Induced Apoptosis
Human Hepatoma PLC/PRF/5 (CD95-negative) cells Phenylethyl Isothiocyanate (PEITC) apoptosis, blocked by NAC and Vitamin E
Cancer Cells Know How to Protect Themselves
My Summary of Cancer Cell Types Killed by EMODs (ROS) (From Gupta et al, 2011)
MORTALITY AND EXCESS ANTIOXIDANTS
Food for Thought
Even Multivitamins are Being Questioned
The Downside of Antioxidants
ANTIOXIDANT FACTS
Some of my books are available on www.amazon.com
MORTALITY
Antioxidant Supplements and Mortality
SUICIDE
Harman's Lamentable Free Radical Theory
Antioxidants Increased the Risk of the Following Cancers
One hundred eighteen (118) Studies Showing Antioxidant Ineffectiveness in Cancer Treatment and thirty-nine (39) Studies Showing Harmful Effects
Suicide and Self-Harm
Reactive Oxygen Species and the Free Radical Theory of Aging
The Role of Reactive Oxygen Species and Mitochondria in Aging
A Midlife Crisis for the Mitochondrial Free Radical Theory of Aging Antioxidants
The National Institutes of Health's Aging Intervention Testing Study Conclusions
New Challenges to Study Heterogeneity in Cancer Redox Metabolism
Antioxidants: Pro or Anti?
Antioxidants Can Even Kill You
Deaths in Japan: Clioquinol
Tragic Deaths of 38 Infants by Lethal IV Antioxidant Vitamin E
The Arrival of Antioxidants
Dr. Howes' Antioxidant Hype Update 2017
Antioxidant Supplements are Everywhere
False Claims Exposed
Americans Will Swallow Anything

Section Two: Cancer and XS Antioxidants

Antioxidants Prove Troublesome Against Cancer
Basic Cancer Considerations
THE FREE RADICAL FALLACY
Diseases Related to Oxygen-Derived Free Radicals: An Erroneous Concept
My EMOD Insufficiency Theory for the Allowance of Tumorous Growths

Immune System Can Halt Growth of a Cancerous Tumor Without Killing It
My Basic Observation on the Lack of Toxicity of Oxygen Free Radicals
Howes EMOD Cancer Therapy
New Drug May Make Tumors Self-Destruct
EMODs Kill Cancer
Carcinostatic Activity of Inorganic Selenium Compounds Due to EMODs
Selenium Properties
Selenium Superoxide Production
Introduction to Low Oxygen Toxicity
High Levels of Antibodies - Low Levels of Cancer?
Inverse Relationship of Allergies and Cancer
Via the Stomach (Food Allergies)
Inhibit Mitochondrial Respiration, Increase EMODs and Kill Cancer
SOD as Possible Target for Killing Cancer Cells
A Super Way to Kill Cancer Cells?
HIV Infection Raises Lung Cancer Risk
My Take on Cancer
Does Telomerase Affect EMOD Production?
Ascorbic Acid Tumoricidal Effects Via H_2O_2
Cheap, Safe Drug Dichloroacetate (DCA), Kills Most Cancers
Dichloroacetate (DCA) Produces Superoxide Anion
Dichloroacetate (DCA) Superoxide Production Blocked by Antioxidant, Ellagic Acid
Free Radical Theory of Aging Unsupported by Human Study
High EMODs and Low Toxicity in Birds
Antioxidants Increased the Risk of the Following Cancers
SUMMARY of Antioxidants Causing Various Cancers
26 STUDIES SHOWING RELATIONSHIP OF HARMFUL EFFECTS OF
ANTIOXIDANTS TO CANCER
Vitamin E and N-acetylcysteine (NAC) Can Fuel the Growth of Lung Cancers

The Simplicity of My EMOD Induced Apoptotic Approach
The State of Cancer: Are We Close to a Cure?
          Boosting the Immune System's 'Arsenal'
          Therapeutic Viruses and Innovative 'Vaccines'
          The Nanoparticle Revolution
          Cancer Treatments and Epigenetics
What are Trends in Cancer Mortality Rates?
          Clinical Context
          Study Synopsis and Perspective
          Study Synopsis and Perspective
          Liver Cancer Continues to "Increase Rapidly"
          Survival Rates
          Study Highlights
          Clinical Implications
Study: Those High-Priced Antioxidants May Be Killing You
Dietary Antioxidants and Human Cancer

INTRODUCTION
ANTIOXIDANTS DURING CANCER TREATMENT
ANTIOXIDANT FOODS AND CANCER PREVENTION
ANTIOXIDANT SUPPLEMENTS AND CANCER PREVENTION
DIETARY ANTIOXIDANTS: CONFRONTING THE DILEMMA
DISCUSSION
A History of Mistakes
    Introduction
    Physiological Effects of Reactive Species
    Pro-Oxidant Effects of Antioxidants
    Clinical Trials with Antioxidants
    Beneficial Effects of Antioxidants
    Antioxidants Do Not Change the Evolution of Diseases Related to Oxidative Stress
    Harmful Effects of Antioxidants
    Conclusions
Antioxidant Supplements Not Beneficial in Cancer
High-Dose Vitamin C Supplements Diminish the Benefits of Exercise

CANCER
EMODs and Spontaneous Cancer Regression

Chemotherapies that Produce Free Radicals, EMODs (Partial List)
Prooxidant Cancer Therapies
Antioxidants May Make Cancer Worse
Good Advice
HOWES' NEW RULES
Spontaneous Regression of Cancer

67 Antioxidant Studies Linked to Increased Cancer Risk –
    A Distillation in Chronological Order: 1986-2014
A List of 41 U.S. Food and Drug Administration-Approved <u>Nontargeted</u> Anticancer Drugs
That Work Through Generation of Reactive Oxygen Species (EMODs):
A List of 40 U.S. Food and Drug Administration-Approved <u>Targeted</u> Anticancer Drugs
That Work Through Generation of Reactive Oxygen Species (EMODs)
My Summary of Cancer Cell Types Killed by EMODs
Heart Disease and XS Antioxidants
Mortality in Randomized Trials of Antioxidant Supplements for Primary and Secondary
Prevention
Antioxidant Profile and Early Outcome in Stroke Patients
T Cells and Reactive Oxygen Species (EMODs)
    Introduction to ROS (EMODs)
    T cells and Oxidative Stress
    ROS (EMODs) as Regulators of Activation-Induced Cell Death
    Conclusions and Perspectives
Reactive Oxygen Species (EMODs) in Cancer
The Naked Mole-Rat

Oxygen Utilization

Thermoregulation

Cancer Resistance

Lifespan

Habitat

Naked Mole-Rat Genome

PROOF EMODs (ROS) ARE AN ESSENTIAL PART OF CANCER KILL

Howes' Rational for EMOD Benefits and Antioxidant Harm

EMODs - A Primary Means of Killing Cancer

Summary of Chemotherapeutic Agents Which Generate EMODs

EMODs, Antioxidants and Apoptosis: All Antineoplastic Drugs Yield EMODs

Radiotherapy Produces Extensive Amounts of EMODs

Photodynamic Therapy (PDT)

Epilogue

Conclusion

Investigators have considered all sorts of excuses for the failures of antioxidants but have rarely, if at all, considered the fact that the Free Radi-Crap theory is "radically wrong."

To accept the teachings and concept of the Free Radi-Crap theory, along with that of oxidative stress, is a "radical" mistake. 5/5/06 RMH

Prologue

 ## My Story in a Nut Sack:

Fundamental Value of Creative Ideas

In evaluating the fundamental value of an individual's creative ideas, one must not rely solely on the preconceived acceptance of the so called good-old-boy network of experts, how many meetings the author has attended, how many organizations he belongs to, how many degrees he has accumulated, how many articles he has published or his demonstrated skills as a medico/scientific politician and con artist. Instead, I suggest that before accepting or rejecting another's new and creative ideas, there is an absolute requirement for the unbiased evaluation of the intellectual integrity of the ideas themselves as regards their inherent honesty, ascertaining their persuasive strength based on the magnitude of prior established data and the courage to judge for yourself. Anything less is cowardly pandering and appeasement of peers, which is likely to stall scientific progress, just as it has done so many times in the past. Do not stick with a popular but erroneous notion just because it is the convenient and comfortable thing to do. Do not be afraid to open your mind and to step into the future of scientific discovery.

My opinions are biased, based upon decades of study and observations and not trammeled by adherence to any outside pressure constraints, marketed products, scientific social clubs, pharmaceutical companies, university faculties or peer groups. I readily admit my biases.

The arguments set forward in my books suggest possible mechanisms for the development of a new paradigm regarding electronically modified oxygen derivatives (EMODs), as they relate to living/breathing aerobic cells.

Oxygen free radical biochemistry, especially that of EMODs, is a gargantuan hodgepodge of confusing and conflicting data and terminology. I have straightened out this colossal mess.

Actually, **free radicals are at the very heart of our energy producing pathways**, which consists of the electron transport chain and oxidative phosphorylation. **Free radicals and EMODs provide the energy of life** due to their reactions involving oxygen.

Our immune system affords us protection for invaders with the use of EMODs.

I have proposed a hypothetical scheme to explain the biochemistry of oxygen metabolism in a unified theory and have proposed an innovative approach based on published literature, experimental data, logic and observations. I have painstakingly assembled the data to form the basis for my hypothesis and to indicate the likelihood that it is valid in vivo.

I am constantly amazed at the inherent intelligence of the normally functioning human body. It is a combinatorial system of interlinked pathways which nourish, sustain and support its attempts at prolonging its existence. It extracts needed sustenance from the environment and the gastrointestinal tract. Basically, our body's only means of intake is via breathing, eating and drinking. It is constantly monitoring the biochemical parameters within itself necessary to prevent disease and extend its life span. We have very little conscious control or input over a vast number of biochemical events which are crucial for our very existence.

These biochemical systems are the end result of billions of years of evolutionary screening and selection. We are the living/breathing culmination of eons of evolutionary wisdom. The inherent sagacity of this selection process is, indeed, mind boggling.

For the life of me, I cannot see how theorists could propose that the generation of electronically modified oxygen derivatives (EMODs) must have been designed as a method of injuring or destroying the body. This idea is counterintuitive to everything that we know about Darwinian development.

Just as our fibrinolytic system must work constantly to prevent intravascular coagulation, our cells must constantly generate sufficient quantities of EMODs to stave off disease and to maintain a state of health and homeostasis. The process of continual EMOD generation is essential for our sustained existence and not the result of a biochemical "leak" with only pernicious intent.

Due to our present ignorance of EMOD functionality, we are just beginning to appreciate their role in the magical process called life. Additionally, belief in the flawed free radical theory, has held hostage progress in enlightenment in their crucial participation in sustaining the vitality of the living/breathing cell.

Hopefully, this time of intellectual stasis is coming to an end. Understanding of EMODs offers us nearly unlimited possibilities for controlling disease and delaying the ringing of the apoptotic death knell of aerobic cells.

My views are skewed to reflect my basic belief that EMODs are of low toxicity and of high benefit to all aerobic cells. Thus, I have written many "selective" books reviewing varying diseases and their relationship to EMOD metabolism. Also, please keep in mind that most diseases have multiple causes, manifestations and courses.

Over the past twenty years my views on disease and health have dramatically changed. Previously, I thought that we existed in a state of health and at some point, diseases such as cancer, atherosclerosis, diabetes and even death came into our lives in the mid to latter decades of life. My views have changed significantly.

Health is maintained by the constant and continual efforts of the body to hold disease in abeyance. If and when the body becomes lax or incapable of fighting disease, then disease manifests itself.

Currently, I believe that we normally live in a zone of homeostasis that is represented by normal blood and urine chemistry values, normal chest x-rays, normal EKG's, etc. However, any of these values are subject to gradual or instantaneous change. Also, we must be fortunate enough or lucky enough to avoid massive trauma or accidental injury induced death. We are subject to constant, complex, internal biochemical monitoring and adjustments.

Diseases or biochemical circumstances collectively change these values and we drift or plunge out of this homeostatic zone of wellness into a condition of illness or if the values change drastically enough, we die. As physicians, we try desperately and usually in a futile manner, to bring these values of normalcy back into kilter. When diseases are combined or when multiple physiological systems are involved, their effects can be more rapid and more devastating as is evident in multi-organ failure.

Our bodies are filled with synchronized biochemical complexities that are actually and presently beyond comprehension.

Sure, we have an elementary grasp on a few of these but the manner in which they are simultaneously choreographed is beyond our grasp. We have compartmentalization of vast quantities of varying chemicals, such as glucose, sodium, potassium, insulin, thyroxin, etc. This compartmentalization and spatial organization occurs from the cell, to the tissue, to the organ level.

Furthermore, our bodies are filled with sensors to monitor each of these agents and to make simultaneous adjustments accordingly, to keep us in a state of health or homeostasis. If the level of these agents drifts from the range of normalcy, disease is manifested and may progress rapidly to death.

Oxygen and EMOD levels are crucial components of this homeostatic model of health.

Wellness is maintained only by the vigilance of an extremely complex intrinsic continuously operational oxidative defensive system. In other words, we are well or alive only because we maintain the ability to constantly fight off or repel pathogens or neoplasia, which can disrupt homeostasis and push us out of our zone of wellness, although it is obvious that any of these chemical balancing acts can go askew.

Our bodies harbor more than ten microbes per cell. Yet, we are not constantly "infected." We are sanctuaries for countless "commensural" viruses and bacteria, which must be persistently held in abeyance. However, they can manifest their pathogenicity the very instant that our bodies become "run down" or immunosuppressed or caught off guard. I believe that this can also be brought on by being in an EMOD insufficiency state.

Apparently, if we exclude accidents or injury, we usually die because of the entrance into our lives of singular or combinations of diseases, which collectively push us farther out of our zone of homeostasis. However, I believe that we can die from the consequences of altered biochemical parameters as a result of these diseases.

This does not have much effect on lifespan because longevity appears to be determined by other more rudimentary factors. In other words, the occurrence of disease can shorten ones genetically determined lifespan. Apparently, many creatures die from "old age." Yet, the genetic limits to death can be rapidly over ridden or accelerated by removal or stoppage of oxygen, water or nutrients, any of which, prevent us from maintaining our zone of wellness.

Even the predatory cats of the African plains instinctively know that by cutting off the oxygen supply of their prey that death will rapidly ensue. This act alone overrides any genetic determinants establishing life span. For aerobes, anoxia equals death and hypoxia portends conditions which "allow" the manifestation of disease. I believe that the EMODs are the primary agents which offer continual prooxidant protection from diseases.

Our bodies must be constantly vigilant in repairing or removing damaged cellular organelles, cellular biochemicals, and their respective tissue and organ counterparts. If not, we "allow" disease to manifest itself and/or to progress.

Obviously, we can be born with or acquire imbalances in any or all of these homeostatic agents and systems. Further, these systems have differing levels of significance and impact upon health. I believe that the crucial EMOD system is of low toxicity, as is evidenced by their constant presence in high numbers.

Please remember that in aerobic life, EMODs are ubiquitous and omnipresent!

Today, we have increased the lifespan because we have prevented or partially controlled some of the more common diseases such as, infections and diabetes and we are avoiding accidental or traumatic death. Lifespans, that are being reached now, frequently represent its "normal" potential for length.

Of particular interest to me is that our increased life spans have been obtained in an era of high generation of oxygen free radicals into the environment (smoke, pollution, increased UV, decreased ozone layer, etc.).

I believe that these EMODs are the fundamental basic elements for prooxidant protection from pathogens and neoplasia and that they are responsible for oxidative self-healing and for staving off death. They keep countless pathogens and neoplastic cells continually in check and they await bodily injury, such that they can assist in collagen synthesis and wound healing. **Just as the body has the fibrinolytic system which continually keeps blood from clotting intravascularly, the body has oxidative systems which keep cancer from growing, plaques from forming, bacteria from becoming pathogenic, herpetic lesions from developing and cataract proteins from coalescing.** My review of the world literature revealed that spontaneous regression is seen in arteriosclerosis, cataract formation, diabetes, a wide range of cancers, HIV/AIDS, arthritis, etc. This gives strong support to my position that we must have in place an intrinsic system to hold in abeyance or reverse these conditions or they would remain or progressively get worse, i.e., spontaneous regression would not be possible.

According to Lester Packer, "A free radical is a molecule with an unpaired electron, seeking to strip an electron from another molecule, and thus having the capacity to damage vital compounds such as lipids and proteins. An antioxidant is basically an electron donor: it can quickly "disarm" a free radical by easily giving up one of its electrons. But in the process, the antioxidant itself becomes a weak free radical. Fortunately, if other antioxidants are present, the original electron donor can be "regenerated," or restored to its antioxidant status." Here, Packer acknowledges the fact that there are varying strengths to free radicals and this is the prime reason that each EMOD species should be discussed on the basis of its specific biochemical characteristics. **Some radicals such as superoxide are puny; whereas, others such as peroxynitrite are powerful**. Unfortunately, and erroneously, free radicals have been labeled as harmful by definition. This is not true, and neither is the blanket statement that all antioxidants are healthful by definition.

Thus, it behooves us to maintain our oxidative abilities and avoid anything which can potentially minimize prooxidant protection. We have repeatedly seen the failure of antioxidants to prevent or reverse diseases and have seen their harmful potential. Still, the radico-phobes deny an overwhelming accumulation of data which discounts and invalidates the free radical theory. They only see through radical-colored glasses, as it relates to disease and aging, which taints their view of biochemical truth.

Randolph M. Howes, M.D., Ph.D.

I believe that just as one disease lowers the immune system and predisposes to another disease that lowering of the EMOD level also predisposes to the occurrence of another disease. They can have a cumulative EMOD lowering effect and produce the "disease clustering" I discuss.

*First a new theory is attacked as absurd; then it is
admitted to be true, but obvious and insignificant;
finally it is seen to be so important that its adversaries
claim they themselves discovered it.*
William James, *Pragmatism*, 1907

Section One: Mortality and Suicide
Antioxidant Skullduggery

This book pulls together all the published evidence and demonstrates beyond doubt that it is important to avoid intake of excessive amounts of antioxidants.

Pushing Excessive Antioxidants is Manslaughter
Excessive Antioxidant Intake Is Linked to Increased Mortality

Personal Introductory Comments

Our environment is effectively saturated with electronically modified oxygen derivatives (EMODs). Singlet oxygen, a name given to several higher energy species of molecular oxygen in which all the electron spins are paired, is much more reactive towards common organic molecules.

In nature, singlet oxygen is commonly formed from water during photosynthesis, using the energy of sunlight. It is also produced by the immune system as a source of active oxygen. The air and the water are fundamentally filled with hydrogen peroxide, superoxide anion, the hydroxyl radical and singlet oxygen. **Yet, the birds and the bees have not fallen from the sky, the aquatic dwellers are not floating belly-up and the surface occupiers are upright and flourishing.**

The dreaded **"oxypocalypse"** has not materialized as predicted by the free radical theorists. These free radical preachers of doom and gloom have invalidated their own brainwashed theory by its inherent lack of predictability and its inconsistencies.

I am convinced that a state of health is due to the constant warding off of disease. This is not particularly a new general concept but my proposal for the salutary role of EMODs is uniquely new. These diseases are frequently the result of bacteria, viruses, fungi, parasites and neoplasia. Shielding us from disease is, in part, accomplished by the "continuously operational intrinsic oxidative defensive system."

I have racked my brain to find the most egregious examples for making my case that EMODs are beneficial essentials and are, in fact, of low toxicity. On the one hand, patients with chronic granulomatous disease (CGD) cannot produce superoxide anions and they are subject to repeated infections and tumors called granulomas. These CGD conditions ultimately culminate in their early deaths. Further, Russian studies have shown that rats and mice raised in a superoxide-free atmosphere will die in 2-3 weeks. Additionally, knock out mice for the gene which makes superoxide dismutase (SOD) quickly die. Please remember that SOD functions to produce H_2O_2 from superoxide. Swiss acatalasemics, who cannot produce catalase to break down H_2O_2, live essentially normal lives. Thus, I ask, "How toxic is H_2O_2 ?"

Myeloperoxidase (MPO) serves an unexpected, protective function in atherosclerosis in a hypercholesterolemic murine model and it generates hypochlorous acid (HOCl). The free radical theory states that EMODs are causal for atherosclerosis but decreased levels of superoxide do not prevent atherosclerosis.

Additionally, **individuals with total MPO** deficiency show a high incidence of malignant tumors. These facts are strong arguments and support for my Unified theory and they serve as powerful arguments against the free radical theory.

Adipocytes of obese patients exhibit inflammation and hypoxia and have limited capability to generate EMODs. Obese patients are subject to increased risk of infections, delayed wound healing, diabetes, cancer and cataract formation. Fat cells produce estrogen-like antioxidants in abundance.

Obviously, I believe that these circumstances are directly related to decreased EMOD generation and to the development of a condition that I call **"Howes' EMOD insufficiency syndrome."** **(The ROSI syndrome of Howes)**

Immunosuppressed patients are subject to infections and cancer. In 2001, 62,000 people in the United States died of influenza or pneumonia. And each year, around 215,000 people in the United States die from severe sepsis, which is more than die from breast, colorectal, pancreatic, and prostate cancer combined. Tuberculosis, once considered under control, now kills 1.7 million every year.

At the same time, new infectious diseases are emerging around the globe in such forms as bird flu and severe acute respiratory syndrome (SARS). I believe that these patients have an "endogenous EMOD deficiency state." (Reactive Oxygen Species Insufficiency Epidemic syndrome, **The ROSI syndrome of Howes**)

We now know that all antibody immune reactions involve the participation of EMODs, such as H_2O_2, hypochlorous acid, ozone and singlet oxygen. Prooxidant protection is crucial in combating bacterial, fungal, viral, and protozoan pathogens. HIV/AIDS patients suffer from repeated infections and Kaposi's sarcoma. These patients have a lacking ability to generate EMODs. Please remember that diabetes and arthritis are considered to be associated with immunosuppression by many authors.

There is no doubt that oxidation represents our first line of defense against pathogens and neoplasia. We know that the proper healing of wounds requires the participation of EMOD generating enzymes and without adequate collagen synthesis, normal healing cannot occur. This is an obvious example of oxidative self-healing.

On the other hand, hyperbilirubinemic babies treated with phototherapy, which specifically generates excited singlet oxygen, are not only cured of their jaundice but they live normal lives and attain normal life spans. These babies have been exposed to EMODs at one of the most vulnerable times of their lives (they have low antioxidant enzyme levels, their diet does not contain many of the low molecular weight antioxidants and colostrum is high in hydrogen peroxide), yet, they suffer no ill effects of these high doses of singlet oxygen.

Free radical theorists predicted that combining iron and peroxide would produce the deleterious and possibly mutagenic hydroxyl radical, via the Fenton reaction, resulting in damaging carcinogenesis because of uncontrolled reactions of the hydroxyl radical with proteins and DNA. Countless patients have irrigated, flushed, washed, and debrided wounds with 3% H_2O_2 for decades and by doing so have directly combined H_2O_2 with hemoglobin-iron at the injury or bleeding site.

According to the free radical theory, this combination of H_2O_2 and iron should predictably produce the mutagenic and damaging hydroxyl radical, but I cannot find even one example of tumors or cancers at these sites in the literature. In fact, all I can find are examples of H_2O_2 aiding in the killing of pathogens and augmenting wound healing.

Exercise has proven to be of great overall benefit to generalized health and to serve as an adjunct in the treatment of conditions considered to be caused by high levels of EMODs, such as diabetes, cancer, atherosclerosis, obesity and arthritis. Unquestionably, exercise always results in generation of high levels of EMODs and they are associated with salutary effects, reduction of diseases and their comorbidity and not related to harmful sequelae.

In my sixth book, I was only going to discuss and review diabetes, but it readily became apparent that I would have to include discussions of obesity, hypertension and atherosclerosis as well, since they are intimately linked to the diabetic condition. Additionally, **obesity has been linked to chronic diseases, such as type 2 diabetes, cardiovascular disease, hypertension and stroke, and some forms of cancer.**

Further, I wished to present an overview of my electronically modified oxygen derivative **EMOD insufficiency syndrome, which includes cancer, atherosclerosis, strokes, obesity, diabetes, arthritis, hypertension, premature aging and cataracts**. Many of these diseases serve as risk factors for others in this group. **Having a risk factor for a disease makes the chances of getting a condition higher but does not always lead to development of the disease. Also, the absence of any risk factors or having a protective factor does not necessarily guard you against getting the disease. There are other cases where one disease raises the risk of another.**

Major trials do not prove the value of antioxidant vitamins. In fact, many have failed. They suggest that many of the so-called antioxidant benefits may have been overstated and that there may be some serious adverse effects. Currently, clinical trial results are our best bet to assess the risks and benefits of antioxidant supplements. However, although positive responses have been reported with these agents, many reports have shown no benefit, and there are widespread disparities in the literature.

My work is not going to be just a step forward, it's going to be a potential revolution. Once we have identified the mechanism for disease causation, steps can be logically taken for control or cure of these disease entities.

Ninety nine percent of today's medical and scientific papers assume or accept the validity of the free radical theory and the concept of so-called oxidative stress. The literature is filled with erroneous prejudice towards EMODs. Investigators design their studies to validate the erroneous Free Radi-Crap theory and when their results fail to do so, they seriously question their very own experiments.

Benjamin Franklin had stated in the Declaration of Independence, "We hold these truths to be self-evident." Franklin had borrowed this term from Newton. Scientists, including Einstein, had tried to describe natural phenomenon with theories of unification. I have done similarly with the biological systems which utilize oxygen metabolism and I believe that I have presented facts of oxygen metabolism which are self-evident, with regards to its essentiality and its low toxicity, even though this contradicts many in vitro experiments. The magic occurs in the living/breathing cell. In vitro experiments are filled with dead artifacts.

My work is bolstered by the application of **Occam's razor or the rule of parsimony**, in that the simplest explanation is usually the right answer. The most basic aspect of disease manifestation is alterations in oxygen and EMOD levels. Oxygen metabolism is a key factor in disease causation, homeostasis, prooxidant protection and oxidative self-healing. It is crucial in the maintenance of a continually operational intrinsic oxidative defensive system.

I have attempted to keep abstracts in a recognizable form and thus have not labeled them as such. Also, I have tried to be certain to reference all materials cited in my books.

If one were to take the literature on oxygen free radicals (EMODs) in its entirety, one would probably be led to discount all theories since results are spread all over the place and filled with inconsistencies.

The widely varying and confusing risk measurements from different studies raise a question as to a proper risk value for diabetes and CVD. Knowledge about the key mechanisms that cause a disease is a usual requirement for designing methods to reduce or eliminate its risk.

It has been said that one must use caution in recommending antioxidant supplements to people with diabetes. One basis for this caution is that, under certain circumstances, vitamin E or C can actually act as a prooxidant. Also, vitamin C shares several cellular transport mechanisms with glucose, and it can increase the rate of absorption of iron, which is a prooxidant. However, because of decades of study, I do not believe that this prooxidant activity is as harmful as is their antioxidant activity.

A lack of oxygenation leads to a lack of oxidation by EMODs, which leads to or "allows" for disease manifestation. Ground state oxygen lacks the reactivity necessary to participate in the complex biochemical reactions required to maintain homeostasis, to protect us against pathogens and to sustain our health. EMODs are sufficiently reactive to protect and to heal us.

Additionally, each EMOD species must be considered separately due to differences in reaction characteristics and chemical qualities. We can no longer lazily and incorrectly lump them under one erroneous and misleading heading, such as reactive oxygen species.

The use of the term "oxidative stress" is misleading and overlooks the fact that oxygen species are in a state of constant change in the organelles, the cell itself, the tissue and the entire aerobic organism. Our redox status is a changing entity.

As I said in my book entitled, *Medical and Scientific Significance of Oxygen Free Radical Metabolism*, "The primary process which stands between us and constant or perpetual infections or manifestations of neoplasia, is our ability to generate oxidative events in the form of EMODs and electron and proton movement, which form a major part of our "intrinsic defensive system."

Cyclic neutropenia, chronic granulomatous disease and spontaneous regression of cancer illustrate my point. Neutrophils produce pathogen-protective EMODs. It appears that the key players, among others, in the immune system, in preventing infections and cancer, are antibodies, macrophages, lymphocytes, neutrophils, T-cells and Interleukin-2.

All of these can and do generate or stimulate EMOD production.

Antibodies can generate hydrogen peroxide (H_2O_2) from singlet molecular oxygen ($^1O_2^*$). Antibodies produce up to 500 mole equivalents of H_2O_2 from $^1O_2^*$, without a reduction in rate. There is an enormous potential for H_2O_2 production by antibodies. **I believe that this inter-relationship between $^1O_2^*$, H_2O_2 and O_3 will prove to have great significance in the generation of a DOA (deadly oxidative assault) and prooxidant protection towards cancer and disease.**

Oxygen plays a pivotal role in the maintenance of life for all eukaryotes, with the exception of strict anaerobes. Eukaryotes have developed mechanisms to sense and respond to decreased oxygen levels. **All living aerobic organisms have adaptive mechanisms to maintain oxygen homeostasis.**

Antioxidant vitamins demonstrate how a therapy that is reportedly safe, can do damage.

People say antioxidant vitamin ingestion cannot do any harm but when it is being promoted for cancer, heart disease and stroke cures, then there is a serious problem. Making false claims about treating colds is one thing but it is quite another to make false claims about reversal of aging or curing cancer.

Newspaper or television interviews and advertisements, concerning antioxidant vitamins, contain statements that are far more positive than warranted by clinical trial study results. Frequently, benefits from these antioxidant vitamins in these reports or advertisements are overstated.

Relative to statin drugs, a systematic search of the Cochrane Database of Systematic Reviews, considered by many to be the most objective medical science reporting of all, showed that all of the industry-funded meta-analyses of drugs recommended the experimental drug without reservations, while none of the Cochrane reviews did so, even though the estimated treatment effects were the same in both cases. (Jorgensen, Gotzsche, 2006).

Peter C. Gotzsche at the Nordic Cochrane Center in Copenhagen, a coauthor of the meta-analyses report, said in an interview that he would now ignore any meta-analyses funded by drug companies. (Anon. Upfront: calling the tune. New Scientist 2006;192(2573):6).

Experts urge caution with dietary supplements because anticipated benefits of antioxidant vitamin use, once widely thought to prevent cancer, heart disease and strokes, have failed to live up to their expectations when put to rigorous testing.

"Developing treatments that attack cancer cells but leave healthy tissue unharmed is the holy grail of cancer research." (Josephine Querido, Cancer Research UK. Senior science information officer).

Many tell you that antioxidant vitamins are completely safe, but will you bet your life on it?

My Opinion on EMODs and Antioxidant Vitamins

For decades, reports have praised the benefits of the antioxidant vitamins and assailed the role of EMODs. Investigators and marketers attempted to stack up enough anecdotal maybes that they hoped to create the illusion of a scientific definite but recent randomized controlled trial have blown these charlatans out of the waters of scientific truth. Yet, they continue to swim in the shallows of ignorance and the dangerous currents of denial.

In 2007, **Dr. Michael Sporn, past Head of the National Cancer Institute Chemoprevention Laboratory and current professor of pharmacology and toxicology at Dartmouth Medical School and Eminent Scholar, NCI's Center for Cancer Research, said that, "The use of dietary substances, like the antioxidant vitamins C and E, has been pretty much a colossal failure for protection against almost any kind of human disease."**

However, I say that the antioxidant vitamins A and E, and to a lesser extent vitamin C, have performed their functions beautifully (which is not the prevention of cancer, heart disease or strokes) but the free radical theory (FRT) has been a uber-colossal failure.

I can find no diseases, other than routine vitamin deficiencies, which are prevented or reversed by the administration of the antioxidant vitamins. Actually, common diseases and an early demise are increased with their use. The FRT was based on poor and inappropriately applied radiation science and a preponderance of subsequent scientific data has now invalidated the free radical theory.
Yet, the FRT has been so ingrained into current medico-scientific thinking that it continues to bias countless studies and the consequent analysis and interpretation of data. Investigators are reportedly "disappointed" in their results, when these antioxidant vitamins fail to perform as predicted by the FRT and rarely question the flawed theory upon which their predictions were based. **Nutrients and exercise, not antioxidant supplements, should be emphasized for overall health.**

During the evening of 11-05-07 I had two epiphanies: 1) all things age including subatomic particles. We call it "decay" instead of aging in this instance. 2) antioxidant vitamins can be thought of as "precursors to oxidants." They could be considered harmful because they wait around to dump electrons on poor oxidants. It all depends on how one looks at it.

The Howes Doctrine

The data speaks for itself and demonstrates the increased risk of inadvertent maladies and increased mortality caused by antioxidant overkill.

In my past books, I have presented over 500 scientific studies, on approximately 16 million human participants, showing that the antioxidants are no more effective than placebo or dummy pills.

If you approach this data with an open mind, you will be awakened to the "suicidal potential" following the inordinate and injudicious use of antioxidants.

In the absence of a deficiency state, antioxidants are not nearly as beneficial as we thought and not close to being as safe as we had been led to believe, especially when used in excess.

Our foods are being pumped full of antioxidants and even junk foods and drinks are loaded with so-called nutrients (or pro-nutrients, whatever that may be) to give them a false perception of a healthier profile. Thus, the risk of overdosing or getting too much of them is increasing rapidly.

More than $20b is still being spent annually on the antioxidant vitamins A, C, and E with more than 6 million tons of the latter projected to be consumed annually on a global basis. (Parker, 2011)

Fads, such as the beta carotene and the vitamin E craze, really backfired when testing revealed increased health risks with their use. Yet, many continue to gulp them down with gusto.

Shockingly, antioxidant use has been associated with increased risk of cancer, heart disease, strokes, diabetes, sperm damage and a host of other health problems and diseases.

Even worse, many studies have shown an increased risk of dying (mortality) linked to the injudicious use of antioxidants. Additionally, on the negative side, antioxidants are capable of interfering with chemotherapy, radiation therapy and photodynamic therapy. Pharmacological doses of antioxidants will interfere with common cancer treatment modalities.

The truth has been found but it has not been told to the consuming public.

Consumers are confused and are getting angry for being "radically" misled.

Even today, manufacturers, middle men, distributors, sales people and profiteers have formed an "antioxidant conspiracy," which collectively releases half-truths and lies regarding the harmful potential of the antioxidants. Truth is seldom found in their sales pitches or literature. Basically, they hide the negative results.

They have arrived at an agreement or a plot to subvert the truth and to talk only about antioxidants as being "secret cures," "miracle cures," "medical breakthroughs," "scientific wonders," "health sensations" and "chemical marvels of our time."

Currently, no one abuses these terms more frequently than Dr. Mehmet Oz, (and to a lesser degree, Dr. Andrew Weil and Dr. Deepak Chopra). Are they **the "three antioxidant stooges?** Pushing these dangerous products has helped make them very wealthy.

They have intentionally avoided "spilling the scientific beans" showing **the verified unintended deadly consequences** of ingesting inordinate amounts of antioxidants.

Shockingly, three major scientific studies were stopped nearly two years early due to unexpected harmful outcomes and adverse effects for study participants: ATBC, CARET and SELECT. (Heinonen et al, 1994, ATBC) (Omenn et al, 1996, CARET) (Lippman et al, 2009, SELECT) respectively.

The supervisory physicians felt it unethical to continue with these studies because of the obvious harmful effects showing up in those taking the antioxidants. The shutdown of these major studies speaks volumes against the casual use or over use of antioxidants.

This has led to the use of the negative term **"SCAM" therapies** (supplements, complementary, and alternative medicines). The truth must be told.

Polyphenols (including phenolic acids and flavonoids) are the most abundant class of antioxidant phytochemicals contained in fruits and vegetables.

Many phytochemicals (some say 4,000 flavonoids and over 600 carotenoids) can be rapidly metabolized into chemicals other than those with antioxidant properties. These have been and still erroneously are being called "antioxidant miracles."

A research team at the Linus Pauling Institute (LPI) and the European Food Safety Authority state that flavonoids, inside the human body, are of little or no direct antioxidant value. Body conditions are unlike controlled test tube conditions, and the flavonoids are poorly absorbed (less than 5%), with most of what is absorbed being quickly metabolized and excreted.

The increase in antioxidant capacity of blood seen after the consumption of flavonoid-rich foods may not be caused directly by the flavonoids themselves, but is probably due to increased production of uric acid resulting from excretion of flavonoids from the body. According to Balz Frei (Linus Pauling Institute), "We can now follow the activity of flavonoids in the body, and one thing that is clear is that the body sees them as foreign compounds and is trying to get rid of them."

Facts of this nature have led to the controversial results observed in antioxidant and supplement studies.

Looking at all the evidence, it's clear that antioxidant harmful potential has been overlooked or ignored. If there are things we can do to reduce our risk of cancer, heart disease and stroke, we should do as much as we possibly can to avoid them, which amounts to simple preventative measures.

Taking excessive antioxidants is a major risk factor and should be avoided.

The size of the problem is such that it is now a major public health concern. The crime scene has become worldwide because sales of the so-called "protective healthcare" products are increasing in Europe, India and China. The swindle enwraps the globe.

Hundreds of millions of unsuspecting people have fallen for the con.

My research has shown that cancer cells, the malaria parasite and various bacteria have strong antioxidative defenses. The obvious question is, WHY? The obvious answer is for their own self-protection against an oxidative (prooxidant) death. Please keep this in mind.

Also, cancer cells concentrate the antioxidants, lactic acid, vitamin C and glutathione, which I believe is for self-preservation of the malignant cells. Excessive antioxidants in the cancer cells prevent or block an electronically modified oxygen derivative (EMOD) induced cancer cell death.

Insights Gained Through My Research

Also, my research has led me to some new connections. **Down's syndrome patients rarely develop solid tumors, such as those of the breast or lung. They virtually never develop high blood pressure, heart attacks or hardening of the arteries, strokes and certain types of cancer**. Yet, they reportedly have high levels of so-called oxidative stress. They have a trisomy 21 defect, which alters SOD, and increases hydrogen peroxide levels.

I believe that the same thing is true for Huntington's disease, whereby victims have a high oxidative capacity but have low risks for developing cancer.

This appears to me similar to the situation in the naked mole-rat, which also has a very high prooxidant level or oxidative (stress) capacity. There has never been a case of cancer reported in a naked mole-rat and it has the highest known oxidative stress level of all rodents.

I believe that the trisomy 21 of Down's, which have a triplicate amount of the prooxidant superoxide dismutase (SOD) enzyme, produces inordinately high levels of hydrogen peroxide. The hydrogen peroxide serves to protect the Down's from cardiovascular disease and cancer.

These facts are the threads of truth coursing throughout my books. Putting them all together gives unanticipated clarity to the overall data and my UTOPIA theory. This also lends support for my proposals associated with electronically modified oxygen derivative (EMOD) insufficiencies, disease allowance and disease coexistence.

I will discuss seven major oxidative pathways which are **crucial** for homeostasis, health and normal functioning. Any or all of these pathways may be interfered with or completely blocked by excessive antioxidant ingestion.

The seven pathways involve: **cancer protection, pathogen protection, energy production, wound healing, immunity, cell communication and detoxification**.

Antioxidants are nondiscriminatory and act as "dirty bombs," in that they can and do indiscriminately attack simultaneously at many oxidative sites. These multiple attacks make antioxidants especially unpredictable and dangerous.

But, just as with prooxidants, antioxidants have a "pecking order" which determines some of their reactivity.

Also, please remember the dangers of naturally occurring antioxidants, such as **cholesterol, estrogen, testosterone, bilirubin and uric acid**. Hypercholesterolemia is the number one risk factor for heart disease. Estrogen feeds uterine, ovarian and breast cancer. Testosterone feeds prostate cancer. Hyperbilirubinemia can cause brain damage in babies (kernicterus) and hyperuricemia is causative of gout and pathognomonic for cardiovascular disease. Sudden death has been suspected in professional wrestlers, secondary to steroid overuse.

I believe this speaks volumes about the dangers of antioxidants, even those considered to be "natural."

It is really hard to report on this subject and still keep my cool.

I cannot imagine how antioxidant supplement manufacturers can look people in the eye and keep selling them these potentially harmful and provably fatal products.

Significant volumes of scientific data exist to show that ingestion of excessive amounts of antioxidant supplements result in increased rates of mortality, cancer, heart disease and strokes.

Therefore, **anyone who ingests unnecessarily huge levels of antioxidants is, knowingly or unknowingly, committing suicide**. All of these diseases (meaning cancer, heart disease and strokes) account for the leading causes of death in America. And, they have all been scientifically shown to be directly linked to ingestion of excessive levels of antioxidants.

Other parts of the world have to deal with malaria, cholera, and contaminated water supplies but we do not. We have to deal with manmade agents of death, such as drugs, FFA approved medications, guns, abortion, automobiles and supplements.

The unrestrained marketing of antioxidants has created a huge public health problem on a worldwide basis. It is my intent to point out the stupidity of being hoodwinked by clever marketeers, who are only interested in their profit margins.

Maintenance of your health is of paramount importance. So, do not be easily misled by misinformed physicians and intentionally deceptive supplement manufacturers.

Be informed. Be smart and be healthy.

In my prologue to my book, Cancer Killing, Suppression and Protection, I said the following:

Apoptosis or cellular suicide is one of the most important means of eliminating precancerous and cancerous cells from the body. The two-other means are necrosis and autophagy.

In many of my books, and strictly based on the scientific literature, I have emphasized the fact that electronically modified oxygen derivatives (EMODs, formerly called reactive oxygen species, ROS or oxygen free radicals) play a crucial role in the killing (apoptosis) of cancerous cells.

I refer to this as a "radical way to stop cancer."

The primary means that the human body uses to rid itself of precancerous and cancerous cells is mediated by electronically modified oxygen derivative (EMOD) induced apoptosis. This crucial process can be blocked or negated by either small molecular antioxidants or by antioxidant enzymes. Cellular proliferation, cellular arrest and cellular suicide appear to be modulated by relative concentrations of EMODs.

I spite of claims to the contrary, I have also stressed and highlighted the point that so-called "antioxidants" can effectively shield or protect cancer cells from being killed by exposure to EMOD prooxidants.

I believe that curing and preventing cancer revolves around the two-following essential, evidence based, scientific conclusions: EMODs kill cancer cells and antioxidants protect cancer cells. With this knowledge, the time has come to eradicate cancer from our midst and to protect those who have not yet fallen victim to the current global cancer pandemic.

In this book, I present many *in vitro* cases whereby EMODs are proven cancer cell killers. Naturally, one may ask, "Does this concept hold true for *in vivo*, human cancer cells?" The answer is a resounding, "Yes."

The chapters on chemotherapy, radiation therapy, photodynamic therapy, IV megadoses of vitamin C and the Howes Singlet Oxygen tumoricidal system, exercise and the naked mole rat all involve EMOD induced cancer cell apoptosis. Thus, **there are millions of *in vivo* human examples of cancer cell killing and suppression to support the cell culture experiments**.

Many human cancer cells overproduce hydrogen peroxide. High levels (up to 0.5 nmol/hr/10^4 cells) of hydrogen peroxide are constitutively released from a wide range of human tumor cells. I believe that this makes the tumor cell more vulnerable to increases in EMODs and creates selectivity for all modalities of cancer therapy. In other words, adding a specified amount of EMODs to neoplastic and normal cells can induce apoptosis in a cancer cell without doing harm to normal cells.

We must endeavor to apply this rudimentary prooxidant knowledge with due haste.
Prof. R.M. Howes MD, PhD

Preliminary warning:
IN GENERAL,
STUDIES HAVE SHOWN THAT
ANTIOXIDANTS ARE NEITHER TOTALLY SAFE
NOR PREDICTABLY EFFECTIVE.

Physical activity (exercise) is tantamount to
electron activity, whereby triplet oxygen
is empowered, such that it can
achieve its missions
of pathogen and cancer protection
and energy production.
R. M. Howes, M.D., Ph.D.
1-25-15

"Exercise deficiency diseases"
are the same thing as
"EMOD insufficiency diseases"
R. M. Howes, M.D., Ph.D.
1-7-15

Section One: Mortality and Suicide

Decades of overly exuberant antioxidant vitamin claims, regarding disease prevention and antiaging, have not been supported by rigorous scientific testing and negative studies have been largely denied or ignored by the dietary supplement industry.

Myths, half-truths, and outright lies are commonly used to promote their sales since there is minimal governmental oversight of their effectiveness or of their harmful potential. The free radical theory, which served as the basis of antioxidant vitamin studies to prevent disease, lacks predictability, fails to meet the requirements of the scientific method, and has consequently been invalidated.

Antioxidant vitamins have such widespread use that their potential to do harm has become a global public health issue. We must follow the fundamental medical precept of Hippocrates: ''First, do no harm.''

We must separate fact from factitious and ''myths of marketing'' from scientific truths.

Millions of people swear by vitamin supplements. **But many are wasting their time and could even be unintentionally harming themselves; thus, committing indirect suicide.**

Introduction

I ask, "Why would anyone knowingly, and without coercion, ingest excessive levels of antioxidants that are known to cause such a wide range of adverse effects, including increased risk of cancer, heart disease, strokes and your chances of dying?"

In 2007, Bjelakovic et al demonstrated the increased of cancer, heart disease, strokes and overall mortality associated with excessive use of antioxidants. (Bjelakovic et al, 2007).

According to Bjelakovic and Gluud, "Until recently, the available data regarding the adverse effects of dietary supplements has been limited and grossly underreported." High levels of intake (of antioxidants) increase the risk of toxic effects and disease," they wrote.

Regulatory authorities and governmental agencies should only allow safe products on the open market and for wide spread human consumption. Currently, antioxidants undergo no such regulations or standards.

More than $20b is still being spent annually on the antioxidant vitamins A, C, and E with more than 6 million tons of the latter projected to be consumed annually on a global basis. (Parker, 2011)

Bisphenol-A (BPA) is an antioxidant that is found in the lining of canned foods, cash register receipts, dental fillings, some plastics and polycarbonate bottles marked with the number 7. BPA is best known as a hardening agent in plastic bottles and used to line the inside of metal cans to prevent spoilage. Scientific studies have linked BPA at lower levels than those found in the Harvard study to cardiovascular disease, diabetes and obesity in humans.

More than 1 million pounds of BPA are released into the environment each year and most Americans have BPA in their urine.

If the lie, "that antioxidants are wonder drugs", is repeated enough, it becomes accepted as truth. Yet, scientific evidence continues to stack up against the misconception that antioxidant overloads are good for you. In fact, antioxidant overkill is definitely harmful.

One would think that the overwhelming incriminating evidence against antioxidants in my book, "*Dangers of Excessive Antioxidants In Cancer Patients*" would bring a closure to the debate on the use of antioxidants in cancer patients. But, it probably will not.

And, in 2011, overall, 81% of women diagnosed with early-stage breast cancer said they'd used at least one supplement containing antioxidants -- either within multivitamins or in the form of single-vitamin supplements -- in the two years after being diagnosed. Of 89 women who used carotenoids six to seven days per week, 18% died of breast cancer; that compared with just under 7% of women who did not use carotenoid combinations.

Results reported in the journal Cancer found that **women who used any combination of carotenoids had a higher risk of dying from breast cancer, or from any cause.**

New work presented in Nature, 2011 suggests that EMODs might have a role in mitigating (alleviating) certain cancers. Reactive oxygen species (ROS, EMODs) get a bad press, as evidenced by the notable trend in the use of dietary and cosmetic antioxidants.

There are life threatening consequences to antioxidant over loading.

There are many who have been "radically" misled to believe that the frequent ingestion of antioxidants will be helpful in their efforts to maintain optimal health, live longer, and even prevent or cure non-deficiency diseases. But scientific studies show that they have been wrong.

This book is a "selective" world literature review.

The currently accepted tortured logic of the free radical theory teaches that oxygen and its radicals are highly toxic, even lethal.   Actually, superoxide and hydrogen peroxide are ubiquitous and omnipresent in steady state levels in all aerobic cells, especially in post mitotic cells with high oxygen consumption, such as the brain and the heart. Thus, I ask, "How toxic can EMODs be?"

EMODs (ROS) are not as harmful as you might've imagined,
not as destructive as you might've thought
and not the "inner enemy" you might've feared.
In fact, your life depends on them.
R. M. Howes, M.D., Ph.D.
5-28-10

The following was taken from a web blog site on antioxidants.

Saving the best for last: the new book (which I haven't read yet) by oxygen crusader par excellence, Prof Randolph M Howes MD, PhD called "Reactive Oxygen Species vs. Antioxidants: The Oxypocalypse or The War That Never Was". The synopsis alone gives a beautiful smack down of the popular antioxidant crapola:
"Recommended use of antioxidant vitamins to treat varied medical maladies is based on the **invalidated free radical theory. The continued non-acceptance of the null findings of over 500 clinical trials on vitamin and antioxidant supplements has no scientific basis or biochemical plausibility**. The underlying principles of the free radical theory have been proven to have repeated unreliability. It fails to meet the requirements of the scientific method and lacks reproducibility. Yet, the multi-billion-dollar antioxidant supplement industry continues to thrive and trumpet false claims for their **potentially harmful products. No doubt, excessive intake of antioxidant vitamins results in increased risk of heart disease, cancer, strokes and overall mortality**. A working knowledge of redox chemistry is essential to understanding the hundreds of failed trials testing the efficacy of antioxidant supplements and antioxidant vitamins. **Normal redox homeostasis may be pathologically disturbed by overzealous use of antioxidants**. My UTOPIA and ROS insufficiency theories present a new perspective more correctly informed by the most contemporaneous experimental findings and by the most reliable clinical trials and studies. In short, electronically modified oxygen derivatives (EMODs) are essential for human existence and protection. **An EMOD insufficiency "allows" for the development of a multitude of disease entities, including infections, non-healing wounds, infertility, tumor development, arteriosclerotic blockages, and cancer growth and metastasis**. There never has been a "war" between EMODs and antioxidants."
https://thepowerofozone.com/the-antioxidant-scam-part-2-is-the-word-out/

In 2013, an international group of experts led by Eliseo Guallar, MD, Johns Hopkins Bloomberg School of Public Health, Baltimore, Maryland, say they believe that the case for multivitamin use is closed.

"Supplementing the diet of well-nourished adults with (most) mineral or vitamin supplements has no clear benefit and might even be harmful," they write. "These vitamins should not be used for chronic disease prevention. Enough is enough."

What are Free Radicals?

Free radicals are highly reactive chemicals that have the potential to harm cells. They are created when an atom or a molecule (a chemical that has two or more atoms) either gains or loses an electron (a small negatively charged particle found in atoms). Free radicals are formed naturally in the body and play an important role in many normal cellular processes.

So-called free radicals that contain the element oxygen are the most common type of free radicals produced in living tissue. Another imprecise name for them is "reactive oxygen species," or "ROS". Thus, I coined the more accurate term "electronically modified oxygen derivatives," (EMODs).

What are Antioxidants?

Antioxidants are chemicals that interact with free radicals. Antioxidants are also known as "free radical scavengers." Antioxidants interfere with or block the chemical process called "oxidation."

Oxidation is essential for our survival and health homeostasis.

However, oxidation cannot occur if there are not sufficient levels of antioxidants to supply electrons to the oxidizing molecules, such as ground state oxygen. Thus, a certain low level of antioxidants is necessary for our survival and these can be obtained through a well-balanced diet.

Most experts agree that supplements are not necessary unless there is a proven deficiency, or if one lives in an area that is nutritionally lacking, such as Ethiopia or Somalia.

The body makes many of the antioxidants that it uses to neutralize free radicals. These antioxidants are called endogenous antioxidants. However, the body relies on external (exogenous) sources, primarily the diet, to obtain the rest of the antioxidants it needs.

These exogenous antioxidants are commonly called dietary antioxidants. Fruits, vegetables, and grains are rich sources of dietary antioxidants. Some dietary antioxidants are also available as dietary supplements or as chemical antioxidant food preservatives, and this is where danger enters the picture.

Examples of dietary antioxidants include beta-carotene, lycopene, and vitamins A, C, and E (alpha-tocopherol), beta-carotene, pycnogenol, EGCG, lycopene and omega-3 fatty acids.

Many observational studies, including case–control studies and cohort studies, have been conducted to investigate whether the use of dietary antioxidant supplements is associated with reduced risks of cancer in humans. Overall, these studies have yielded mixed results. (Patterson RE, White E, Kristal AR, et al., 1997)

Because observational studies cannot adequately control for biases that might influence study outcomes, the results of any individual observational study must be viewed with caution.

Randomized controlled clinical trials, however, lack most of the biases that limit the reliability of observational studies. Therefore, randomized trials are considered to provide the strongest and most reliable evidence of the benefit and/or harm of a health-related intervention.

The only oxygen we need is
one pint per breath at the rate of
only a minimum of 21,600 times per day.
Other than that….
R. M. Howes, M.D., Ph.D.
9-12-09

Introductory Discussions on Suicide and Mortality

News flash: Take all the antioxidant supplements you desire but do so at your own peril. Get informed and stay informed.

A List of Major Reactive Oxygen Species and Antioxidant Systems in Living Organisms

ROS	Symbol	Antioxidant System Symbol
Radical ROS		**Enzymatic**
Superoxide	O2. -	Superoxide dismutase SOD Hydroxyl
radical	OH	Catalase CAT
Nitric oxide	NO	Glutathione peroxidase GPx
Organic radical	R	Glutathione reductase GR
Peroxyl radical	ROO	GlutathioneS-transferase GST Alkoxyl
radical	RO	Thioredoxin peroxidase TrxPx
Thiyl radical	RS	Thioredoxin reductase TrxR Sulphonyl
radical	ROS	
Thiyl peroxyl radical RSOO		

Non-Radical ROS/RNS

Hydrogen peroxide	H_2O_2
Singlet oxygen	$1\,{}^*O_2$
Ozone (trioxygen)	O_3
Organic hydroperoxide	ROOH
Hypochlorous acid	HOCl
Peroxynitrite	ONOO-

Nonenzymatic

Glutathione GSH
Glutaredoxin Grx
Thioredoxin Trx
Peroxiredoxin Prx
Sulfiredoxin Srx
Phytochemicals
Vitamins A, C, E
Ceruloplasmin

Organizations Not Recommending Antioxidant Vitamins

Contrary to the common impression, major health organizations do not recommend the antioxidant vitamins A, C and E.

THE FOLLOWING LIST PROVIDES THE MAJOR MEDICAL AND SCIENTIFIC ORGANIZATIONS WHICH DO NOT RECOMMEND THE USE OF ANTIOXIDANT VITAMINS

The following either do not recommend antioxidant vitamins or have found inconclusive evidence of their benefit:

- **The U.S. Food and Drug Administration (FDA)**
- **The American Heart Association (AHA)**
- **The American Cancer Society (ACS)**
- **The National Cancer Institute (NCI)**
- **Institute of Medicine of the National Academies**
- **The American College of Cardiology**
- **The American College of Chest Physicians (ACCP)**
- **The American Diabetes Association**
- **The American Academy of Family Physicians**
- **Scientific Statement from the American Heart Association and the American Diabetes Association**
- **The American College of Cardiology/American Heart Association Task Force on Practice Guidelines**
- **United States Preventive Services Task Force (USPSTF)**
- **The American Cancer Society Guidelines on Nutrition and Physical Activity for Cancer Prevention**
- **The Nutrition Committee of the American Heart Association Council on Nutrition, Physical Activity and Metabolism**
- **The AHA Scientific Position of the American Heart Association**
- **The Canadian Task Force on Preventive Health Care (CTFPHC)**
- **Food and Nutrition Board, Institute of Medicine**

- **The Food and Nutrition Board of the National Academy of Sciences**
- **National Academy of Sciences**
- **The 2006 AHA Diet and Lifestyle Recommendations**
- **The Medical Letter**
- **The Oregon Health and Science University**
- **Food Standards Agency/The British Nutrition Foundation (BNF)**
- **Quackwatch**
- **American College of Cardiology Foundation Task Force on Clinical Expert Consensus Documents**
- **National Institutes of Health State-of-the-Science Conference**
- **The American Heart Association Atherosclerosis, Hypertension, and Obesity in Youth Committee, Council of Cardiovascular Disease in the Young, With the Council on Cardiovascular Nursing**
- **The Physicians Health Study**
- **The 2008 VITAmins and Lifestyle (VITAL) study**
- **The Physicians' Health Study II Randomized Controlled Trial**
- **The Swedish Council of Technology Assessment**
- **National Heart Foundation of Australia's Nutrition and Metabolism Advisory Committee**

Although their conclusions are not iron clad, many prestigious scientific organizations have concluded that, "taking antioxidant vitamins - such as vitamins A, C and E - serves no purpose, and in some cases could likely be harmful."

Such a list is rather astounding because broadcast media presents a never-ending cycle of advertisements pushing the wonders of antioxidants and antioxidant vitamins. One would assume that such advertisements would have the backing of major medical and scientific organizations, but that is not the case.

The above 32 conclusions or recommendations are apparently some of the best kept secrets in America, since antioxidants are being fortified or added to a wide spectrum of commercial products including foods, cosmetics, dermatologics, pet products, beverages, energy drinks, energy bars, fruits drinks, fruit juices, chewing gum, shampoos, etc. **Genetic engineers are hurriedly creating "super foods," which will be "antioxidant-rich"** (Howes R.M. 2009, Am J Cosm Surg).

Since readers have been indoctrinated to use the less accurate term, ROS, I have made the two terms, ROS and EMODs, basically interchangeable in this text. But, the term "EMODs" is a more accurate term and does not refer to charge, radicality or reactivity and does include all reactive oxygen species and oxygen free radicals.

Cancer, heart disease and strokes increase man's mortality and should be included in any discussion of the suicidal aspects of antioxidant ingestion. Thus, I excerpted parts of chapter seven from my book entitled, Antioxidants Linked To Deadly Unintended Consequences.

HARMFUL EFFECTS OF ANTIOXIDANTS by Category

Italics indicates harmful effects.

In order to make it easier to find references by category, I present the following. This also makes it easier to link antioxidant harm to specific diseases.

Please remember the following harmful effects of antioxidants. Logically, these increased risks for diseases likely shorten the life span and result in deadly unintended consequences.

Harmful Effects of Antioxidants by Category
MORTALITY (Number of Deaths) (25 Study Reports)
> **CANCER (72 Study Reports)**
> **HEART DISEASE (28 Study Reports)**
> **STROKE (6 Study Reports)**

MORTALITY (Number of Deaths) (25 Study Reports)

1. The β-Carotene and Retinol Efficacy Trial (CARET) (Omenn et al, 1996) (#14,254 heavy smokers and 4,060 asbestos workers) (total #18,314 men and women); *28% increase in lung cancer; 26% increase in CVD (nonsignificant); 17% increase in total mortality among treatment group. This study was stopped 21 months earlier than planned.*

2. Chemoprevention of aerodigestive cancer (Berwick, Schantz, 1997) *Large-scale trials of the anti-oxidant beta carotene have been disappointing; they have shown that among heavy smokers and possibly heavy alcohol consumers, beta carotene increases risk for lung cancer incidence and mortality.* (Berwick, Schantz, 1997)

3. Multivitamin use and mortality in a large prospective study. (Watkins et al, 2000) (#1,063,023 adults) **Multivitamin users had heart disease and cerebrovascular disease mortality risks similar to those of nonusers, whereas combination users had mortality risks that were 15% lower than those of nonusers.** Multivitamin and combination use had minimal effect on cancer mortality overall, *although mortality from all cancers combined was increased among male current smokers who used multivitamins alone or in combination with vitamin A, C, or E.*

4. Women's Angiographic Vitamin and Estrogen (WAVE) Trial (Waters et al, 2002) (#423 postmenopausal women, with at least one 15% to 75% coronary stenosis) In the Women's Angiographic Vitamin and Estrogen Study (WAVE), *postmenopausal women with coronary disease on hormone replacement therapy given vitamin E plus vitamin C had an unexpected significantly higher all-cause mortality rate and a trend for an increased cardiovascular mortality rate.* (Waters et al, 2002)

5. The Medical Research Council/British Heart Foundation **Heart Protection Study**, which is **the study cited by Dr. Gibbons, randomized 10,269 patients to 660 IU/day of vitamin E and 10,267 to placebo control.** *The vitamin E group was associated with about a 10% increase in mortality.* (Gibbons quote, 2002) (MRC/BHF, 2002)

6. Vitamins E & A fail to reduce incidence or mortality of lung cancer: Cochrane Database Syst Rev. 2003. (Caraballoso et al., 2003) (#109,394 participants) When beta-carotene was combined with retinol, data from a single study showed that there was *a statistically significant, increased risk of lung cancer incidence and mortality* in people with risk factors for lung cancer who took both vitamins. (Caraballoso et al., 2003)

7. Cochrane Database Syst Rev. 2004: Vitamins E & A fail to reduce incidence or mortality of gastrointestinal cancer. (Cochrane Database Syst Rev.) (Bjelakovic et al, 2004) (#170,525 participants); *antioxidant supplements significantly increased mortality.* **Beta-carotene and vitamin A and beta-carotene and vitamin E significantly increased mortality,** *while beta-carotene alone only tended to do so.* **Selenium showed significant beneficial effect on gastrointestinal cancer incidences.** *When the selenium trials were excluded, both analyses showed a statistically significant increase in mortality, which was particularly strong in patients taking beta carotene and vitamin A.* CONCLUSIONS: **They could not find evidence that antioxidant supplements prevent gastrointestinal cancers. On the contrary,** *antioxidant supplements seem to increase overall mortality.*

8. Meta-analysis: high-dosage vitamin E supplementation may increase all-cause mortality (Miller et al., 2004) (#135,967 subjects); *high doses of vitamin E increased mortality.*

9. Supplemental vitamin C increase cardiovascular disease risk in women with diabetes (Lee et al, 2004) (#1,923 postmenopausal women who reported being diabetic) *A high vitamin C intake from supplements is associated with an increased risk of cardiovascular disease mortality in postmenopausal women with diabetes.* (Iowa Women's Health study) (Lee et al, 2004)

10. Cochrane Database Syst Rev. 2004: Vitamins E & A fail to reduce incidence or mortality of gastrointestinal cancer. (Cochrane Database Syst Rev. G. Bjelakovic et al, 2004) (#170,525 participants) Antioxidant supplements significantly increased mortality. *Beta-carotene and vitamin A and beta-carotene and vitamin E significantly increased mortality.* **When the selenium trials were excluded,** *both analyses showed a statistically significant increase in mortality, which was particularly strong in patients taking beta carotene and vitamin A.* (Cochrane Database Syst Rev.) (Bjelakovic et al, 2004)

11. Mortality in Randomized Trials of Antioxidant Supplements for Primary and Secondary Prevention; Systematic Review and Meta-analysis (Bjelakovic et al, 2007) (#232,606 participants); *Conservatively, the supplements increase the likelihood of dying by about 5 percent. When looked at separately, they found that Vitamin A increased death risk by 16 per cent, beta carotene by 7 per cent and Vitamin E by 4 per cent.*

12. Risk of Mortality with Vitamin E Supplements: The Cache County Study. (Hayden et al, 2007) *mortality was increased in vitamin E users who had a history of stroke, coronary bypass graft surgery, or myocardial infarction and, independently, in those taking nitrates, warfarin, or diuretics.*

13. Antioxidant supplements for prevention of mortality in healthy participants and patients with various diseases. (Bjelakovic, Nikolova, Gludd, Simonetti and Gludd, 2008 Apr) (#232,550 Cochrane Database Syst Rev.) **Overall, the antioxidant supplements had no significant effect on mortality in a random-effects meta-analysis,** but *significantly increased mortality in a fixed-effect model. Vitamin A, beta-carotene, and vitamin E may increase mortality.*

14. Systematic review: primary and secondary prevention of gastrointestinal cancers with antioxidant supplements. (Bjelakovic, Nikolova, Simonette and Gludd, 2008 Sept) (#211,818 participants) *Antioxidant supplements had no significant effect on mortality in a random-effects model meta-analysis but significantly increased mortality in a fixed-effect model meta-analysis. CONCLUSIONS: There was no evidence that the studied antioxidant supplements prevented gastrointestinal cancers. On the contrary, they seem to increase overall mortality.*

15. Efficacy of Antioxidant Supplementation in Reducing Primary Cancer Incidence and Mortality: Systematic Review and Meta-analysis (Bardia et al, 2008) (#104,196 participants) *Beta carotene supplementation was associated with an increase in the incidence of cancer among smokers and with a trend toward increased cancer mortality.*

16. Total and Cancer Mortality After Supplementation With Vitamins and Minerals: 10 year Follow-up of the Linxian General Population Nutrition Intervention Trial. (Qiao et al, 2009) (#29,584 adult participants) *esophageal cancer deaths increased 14% among those aged 55 years or older. Vitamin A and zinc supplementation was associated with increased total and stroke mortality.*

17. Modification of the effect of vitamin E supplementation on the mortality of male smokers by age and dietary vitamin C. (Hemila and Kaprio, 2009 Apr) (#29,133) The Alpha-Tocopherol, Beta-Carotene Cancer Prevention (ATBC) Study; *Among participants with a dietary vitamin C intake above the median of 90 mg/day, vitamin E increased mortality among those aged 50-62 years by 19%.*

18. No evidence supports vitamin E indiscriminate supplementation (Dotan, et al, 2009, Biofactors) *Major randomized clinical trials yielded disappointing results and recent meta-analyses concluded that indiscriminate, high dose vitamin E supplementation results in increased mortality. The average quality-adjusted life years (QALY) of vitamin E-supplemented individuals was 0.30 QALY (95%CI 0.21 to 0.39) less than that of untreated people.* (Dotan, et al, 2009, Biofactors) (Dotan Y, Lichtenberg D, Pinchuk I. No evidence supports vitamin E indiscriminate supplementation. Biofactors. 2009 Nov-Dec;35(6):469-73)

19. Vitamins and cardiovascular disease (Honarbakhsh, Schachter, 2009) **Randomized controlled trials have not yet supported a role for vitamins in primary or secondary prevention of CVD and have** *in some cases even indicated increased mortality in those with pre-existing late-stage atherosclerosis.* **The trials that used a combination of vitamins that include beta-carotene have been disappointing.** (Honarbakhsh, Schachter, 2009) (Honarbakhsh S, Schachter M. Vitamins and cardiovascular disease. Br J Nutr. 2009 Apr;101(8:1113-31)

20. Antioxidants tied to mixed effects in breast cancer. (Greenlee et al, 2011) *Women who regularly took a mix of carotenoids had a higher risk of dying from breast cancer. Dietary supplements containing high doses carotenoids may be harmful, and people should think twice before taking them."* (Greenlee et al, Cancer 2011)

21. Vitamins May Increase Women's Risk of Dying, Research Finds. (Iowa Women's Health Study, Mursu et al, 2011) *Women who took supplements had, on average, a 2.4 percent increased risk of dying over the course of the 19-year study*, compared with women who didn't take supplements. **In older women, several commonly used dietary vitamin and mineral supplements may be associated with increased total mortality risk;** *this association is strongest with supplemental iron. The use of multivitamins, vitamin B$_6$, folic acid, magnesium, zinc, iron, and copper was individually statistically associated with increased risk of all-cause mortality when compared to nonuse.* (Mursu et al, 2011)

22. Tragic deaths of 38 infants by lethal IV antioxidant vitamin E

In the 1980s, an injectable form of vitamin E, known as E-Ferol, was responsible for the deaths of 38 babies. This was verified by court hearings. Please remember that research has tragically shown that surviving infants who received E-Ferol injections were at an increased lifetime risk for reproductive problems, cervical and vaginal cancer, and other health problems.

Also, please remember the **fatal syndrome characterized by progressive clinical deterioration with unexplained thrombocytopenia, renal dysfunction, cholestasis, and ascites developed in certain infants throughout the United States who had received E-Ferol, an intravenous vitamin E supplement. (THE TRAGIC CASE HISTORY OF INTRAVENOUS VITAMIN E (The New York Times) May 27, 1984 By PHILIP M. BOFFEY).**

Prolonged exposure to E-Ferol was associated with progressive intralobular cholestasis, inflammation of hepatic venules, and extensive sinusoidal veno-occlusion by fibrosis. E-Ferol, contained 25 units per milliliter of dl-alpha-tocopheryl acetate solubilized with 9% polysorbate 80 and 1% polysorbate 20. They proposed that vasculocentric hepatotoxicity is the basis for the observed clinical syndrome that represents the cumulative effect of one or more of the constituents of E-Ferol. (Bove et al, 1985)

All affected infants received E-Ferol; some affected infants received up to 1 ml or more daily. **Both outbreaks ceased shortly after use of E-Ferol was discontinued. Three were jailed for selling the drug (vitamin E) that killed 38 babies.**

The Center for Drug Evaluation and Research, FDA, Rockville, Maryland, concluded that the use of E-Ferol in these neonatal intensive care units was associated with increased morbidity and mortality among exposed infants. (Arrowsmith et al, 1989)

As I mentioned above, "research has shown infants who received E-Ferol injections are at an increased lifetime risk for reproductive problems, cervical and vaginal cancer, and other health problems." This is of grave importance.

Here is another basic point: the increase in disease risk and mortality seen with the antioxidant vitamins cannot be expected to increase the life span. Obviously, they would be expected to decrease or shorten the life span.

Additionally, for those that say that there have been only a "few" negative studies with the antioxidant vitamins, there are now hundreds (over 316) which I have compiled and reported on. That is, indeed, shocking!

23. Also, please remember the work of Drisko et al with IV mega-doses of vitamin C which resulted in 2 deaths and other adverse effects. (Padayatty et al, 2010)

24. Then check out the Japanese deaths from the antioxidant, clioquinol.

Antioxidant vitamin pushers are always saying, "If antioxidant vitamins are so bad, where are the dead bodies?" So, I just listed some of the dead bodies for them!

Since antioxidants increase cancer, heart disease and strokes, it is likely that there are lots of dead bodies attributable to them, but the antioxidants are not identified as the cause of death.

Additionally, the above studies show that the antioxidants increase "overall mortality," which means they increase your chances of dying, which contributes to more dead bodies!

The harmful effects of antioxidant overloading are insidious and not immediate, as with cyanide. But, they are just as real.

25. Who is likely to gain from high dose supplementation of vitamin E? (Lichtenberg, 2011) **The results of the latter prediction were disappointing. Moreover, previous meta-analyses concluded that indiscriminate supplementation of vitamin E at high dose (400 IU or more) results in increased mortality, both cardiovascular and all-cause. *Indiscriminate high dose supplementation of vitamin E is associated with an average loss of about 0.3 QALY.* We cannot recommend high dose vitamin E supplementation.** (Lichtenberg, 2011) (Lichtenberg D. Who is likely to gain from high dose supplementation of vitamin E? Harefuah. 2011 Jan;150(1):37-40)

CANCER (72 Study Reports)

1. Studies on antioxidants: Their carcinogenic and modifying effects on chemical carcinogenesis (Ito et al, 1986). *Squamous-cell carcinomas were induced in the forestomach of rats and hamsters fed BHA* (butylated hydroxyanisole). **BHA and other antioxidants, particularly propyl gallate and ethoxyquin, showed additive effects in inducing forestomach hyperplasia and cytotoxicity. BHA enhanced forestomach carcinogenesis initiated in rats by *N*-methyl-*N'*-nitro-*N*-nitrosoguanidine or *N*-methylnitrosourea (MNU) and enhanced urinary bladder carcinogenesis initiated by MNU or *N*-butyl-*N*-(4-hydroxybutyl) nitrosamine (BBN).** (Ito et al, 1986)

2. Studies on antioxidants: Their carcinogenic and modifying effects on chemical carcinogenesis (Ito et al, 1986). BHT (butylated hydroxytoluene) **promoted urinary bladder carcinogenesis initiated by BBN (*N*-butyl-*N*-(4-hydroxybutyl) nitrosamine) or MNU and thyroid carcinogenesis initiated by MNU (*N*-methylnitrosourea). Ethoxyquin promoted EHEN-initiated kidney carcinogenesis.** (Ito et al, 1986)

3. Pathology of BHA- and BHT-induced lesions (Moch, 1986). Papilloma and squamous-cell carcinoma of the forestomach were increased at the 2.0% level with BHA in dogs. The BHT was fed to Wistar rats at 0, 25, 100 and 250 mg/kg body weight. At the highest dose there was an increase in the number of rats with hepatocellular adenoma and with hepatocellular carcinoma. (Moch, 1986)

4. Alpha-Tocopherol, β-Carotene Cancer Prevention Study (ATBC study) (Heinonen et al, 1994) (#29,133 men); ***50% increase in hemorrhagic stroke deaths*** *among vitamin E group;* ***11% increase in ischemic heart disease deaths among β-carotene group;*** ***18% increase in lung cancer among β-carotene group.***

5. The β-Carotene and Retinol Efficacy Trial (CARET) (Omenn et al, 1996) (#14,254 heavy smokers and 4,060 asbestos workers) (total #18,314 men and women); ***28% increase in lung cancer; 26% increase in CVD*** *(nonsignificant);* ***17% increase in total mortality*** *among treatment group. This study was stopped 21 months earlier than planned.*

6. Energy, nutrient intake and prostate cancer risk: a population-based case-control study in Sweden. (Andersson et al. 1996) (#1,062) *In age-adjusted analyses, there were positive associations of prostate cancer (all stages combined) risk with total energy intake as well as intake of total fat (saturated and monounsaturated), protein, retinol and zinc.* **The positive association with energy intake was stronger for advanced cancer, with an excess risk of 70% for the highest quartile vs. the lowest.** *After adjustment for energy intake, there was no apparent association of prostate cancers (all stages combined) with any of the investigated nutrients. However, a weak positive association between intake of retinol and advanced cancer was observed.*

7. Alpha-Tocopherol and beta-carotene supplements and lung cancer incidence in the alpha-tocopherol, beta-carotene cancer prevention study: effects of base-line characteristics and study compliance. (Albanes et al, 1996) (#29,133 men, smokers) *beta-*

Carotene supplementation was associated with increased lung cancer risk. beta-Carotene supplementation at pharmacologic levels may modestly increase lung cancer incidence in cigarette smokers, and this effect may be associated with heavier smoking and higher alcohol intake. **This study was stopped 21 months earlier than planned. The incidence of lung cancer was 18% higher among men who took the beta-carotene supplement and *eight percent more men in this group died, as compared to those receiving other treatments or placebo.*** (Albanes et al, 1996)

8. Effects of various genotoxins and reproductive toxins in human lymphocytes and sperm in the Comet assay. (Anderson et al, 1997) *Genistein at higher concentrations could also result in DNA damage detectable by either the comet assay or the chromosomal aberrations in both cells in culture and in human sperm and lymphocytes obtained from donors.* (Anderson et al, 1997) (Kulling et al, 1999) (Pool-Zobel et al, 2000)

9. The Nurses' Health Study and Folic Acid and Colon Cancer (Giovannucci et al, 1998) (#88,756 women taking vitamin C and B-carotene, for 8 years); Dr. Andy Ness, of Bristol University, reported in the British Medical Journal in Dec. 2004, that there is *the **possibility of increased risk of breast cancer in women taking folic acid supplements throughout pregnancy**.* The researchers followed up **2,928 pregnant women** who had taken part in a supplemental trial in the 1960s. ***The risk of death from breast cancer was much higher in women who had received high doses of the supplement*** than in those who had been given a placebo.

10. The effects of antioxidant supplementation during Percoll preparation on human sperm DNA integrity (Hughes et al, 1998) (#150 patients) *acetyl cysteine or ascorbate and alpha tocopherol together induced further DNA damage to human sperm*.

11. Estrogenic effects of genistein on the growth of estrogen receptor-positive human breast cancer (MCF-7) cells in vitro and in vivo. (Hsieh et al, 1998)
Tumors were larger in the genistein (750 ppm)-treated group than they were in the negative control group, demonstrating that dietary genistein was able to enhance the growth of MCF-7 cell tumors in vivo. In summary, *genistein can act as an estrogen agonist in vivo and in vitro, resulting in the proliferation of cultured human breast cancer cells (MCF-7) and the induction of pS2 gene expression. Dietary genistein stimulates mammary gland growth and enhances the growth of MCF-7 cell tumors in ovariectomized athymic mice.* (Hsieh et al, 1998)

12. Effects of soy-protein supplementation on epithelial proliferation in the histologically normal human breast. (McMichael-Phillips et al, 1998) *Supplementation studies conducted in humans have raised the possibility that soy may promote breast cancer development. In a trial that randomized women with benign or malignant breast disease to soy supplementation (60 g of soy containing 45 mg of isoflavones) or their normal diet daily for 2 wk, women receiving the soy supplements had increased serum genistein levels in comparison to women on a standard diet, and their histologically normal breast tissue exhibited enhanced breast epithelial cell proliferation and significantly increased progesterone receptor levels.* (McMichael-Phillips et al, 1998)

13. The effect of ascorbate and alpha-tocopherol supplementation in vitro on DNA integrity and hydrogen peroxide-induced DNA damage in human spermatozoa (Donnelly et al, Mutagenesis. 1999) (#Semen samples with normozoospermic and asthenozoospermic profiles (n = 15 for each control and antioxidant group) **Addition of** *both ascorbate and alpha-tocopherol in combination to sperm preparation medium actually-induced DNA damage and intensified the damage induced by H_2O_2.*

14. The phytoestrogens coumoestrol and genistein induce structural chromosomal aberrations in cultured human peripheral blood lymphocytes. (Kulling et al, 1999) *These results, together with previously published reports on the induction of micronuclei and DNA strand breaks in cultured Chinese hamster V79 cells by COUM and GEN, but not DAI, suggest that some but not all phytoestrogens have the potential for genetic toxicity.* (Kulling et al, 1999)

15. Antioxidant vitamin supplements: update of their potential benefits and possible risks. (Maxwell, 1999) **A number of long term, prospective, randomized, placebo-controlled trials examining the protective effect of antioxidant supplements have now been completed. Their results have been generally disappointing and have provided little evidence of efficacy.** *Of greater concern, they have unexpectedly raised concerns that antioxidants, notably beta carotene, might increase the rate of development of cancers in high risk individuals.* **For this reason, regular consumption of antioxidant vitamins supplements cannot yet be advocated as a healthy lifestyle trait.** (Maxwell, 1999)

16. Isoflavonoids (genistein) and lignans have different potentials to modulate oxidative genetic damage in human colon cells. (Pool-Zobel et al, 2000) *Genistein induced DNA breaks in the human tumor cell line HT29 clone 19A* (Pool-Zobel et al, 2000)

17. Genotoxicity of several clinically used topoisomerase II inhibitors (Boos, Stopper, 2000) *Genistein exposure of L5178Y mouse lymphoma cells at concentrations less than 100 nM induced micronuclei formation and mutagenesis at the thymidine kinase locus.* (Boos, Stopper, 2000)

18. Multivitamin use and mortality in a large prospective study. (Watkins et al, 2000) (**#1,063,023 adults**) Multivitamin and combination use had minimal effect on cancer mortality overall, **although mortality from all cancers combined was increased among male current smokers who used multivitamins alone or in combination with vitamin A, C, or E.** (Watkins et al, 2000)

19. Randomized Trial of Supplemental ß-Carotene to Prevent Second Head and Neck Cancer (Mayne et al, 2001) (#264 patients who had been curatively treated for a recent early-stage squamous cell carcinoma of the oral cavity, pharynx, or larynx.); **Supplemental ß-carotene had no significant effect on second head and neck cancer or lung cancer.** Whereas none of the effects were statistically significant, the *point estimates suggested a possible decrease in second head and neck cancer risk but a possible increase in lung cancer risk.*

20. Soy diets containing varying amounts of genistein stimulate growth of estrogen-dependent (MCF-7) tumors in a dose-dependent manner. (Allred et al, 2001) *Genistein supplementation enhanced mammary gland growth and tumor development of estrogen-dependent MCF7 cells in athymic mice in a dose-dependent manner.* (Allred et al, 2001)

21. Mega-dose vitamins and minerals in the treatment of non-metastatic breast cancer: an historical cohort study (Lesperance et al, 2002) (#90 patients with non-metastatic breast cancer who received conventional treatment) *Breast cancer–specific survival (i.e., patients censored only at death from breast cancer) and disease-free survival were shorter in the nutrient-supplemented group* **than in the non-supplemented group, but the differences were not statistically significant.**

22. Selenium and vitamin E supplements for prostate cancer: evidence or embellishment? (Moyad et al. 2002) (# not available) Selenium supplements provided a benefit only for those individuals who had lower levels of baseline plasma selenium. *Other subjects, with normal or higher selenium levels, did not benefit and may have an increased risk for prostate cancer. Vitamin E supplements in higher doses (> or =100 IU) were also associated with a higher risk of aggressive or fatal prostate cancer in nonsmokers from a past prospective study.*

23. Vitamins E & A fail to reduce incidence or mortality of lung cancer: Cochrane Database Syst Rev. 2003. (Caraballoso et al., 2003) (#109,394 participants); *When beta-carotene was combined with retinol, data from a single study showed that there was a statistically significant,* **increased risk of lung cancer incidence and mortality** *in people with risk factors for lung cancer who took both vitamins.*

24. Neoplastic and Antineoplastic Effects of Beta Carotene on Colorectal Adenoma Recurrence: Results of a Randomized Trial (Baron et al, 2003) (#864 subjects who had had an adenoma removed and were polyp-free); *For participants who smoked cigarettes and also drank more than one alcoholic drink per day, beta carotene doubled the risk of adenoma recurrence.*

25. Selenium supplementation and secondary prevention of nonmelanoma skin cancer in a randomized trial. (Duffield-Lillico, 2003) (#1,312). *selenium supplementation was associated with statistically significantly elevated risk of squamous cell carcinoma* and of total nonmelanoma skin cancer. **Results from the Nutritional Prevention of Cancer Trial conducted among individuals at high risk of nonmelanoma skin cancer continue to demonstrate that selenium supplementation is ineffective at preventing basal cell carcinoma and that** *it increases the risk of squamous cell carcinoma and total nonmelanoma skin cancer.*

26. Lycopene increases urokinase receptor and fails to inhibit growth or connexin expression in a metastatically passaged prostate cancer cell line: a brief communication (Forbes et al, 2003) Lycopene *has a potentially unwanted effect of upregulating expression of the urokinase plasminogen activator receptor and facilitating invasion while failing to significantly inhibit proliferation or to induce detectable levels of the gap junctional protein connexin 43 expression.* Our results indicate that **some caution should be taken with regard to**

use of lycopene to treat potentially advanced and metastatic prostate cancers. (Forbes et al, 2003)

27. "Food and Chemical Toxicology"; Effects of Dietary Antioxidants and 2-Amino-3-Methylimidazo[4,5-f]-Quinoline on Preneoplastic Lesions and on Oxidative Damage, Hormonal Status and Detoxification Capacity in the Rat. Long-term administration of the antioxidants lycopene, quercetin and resveratrol *caused precancerous sores in the rats' livers as well as damage to the DNA of certain cells of their immune system.* (Breinholt et al, 2003)

28. Dietary tenistein results in larger MNU-induced, estrogen-dependent mammary tumors following ovariectomy of Sprague-Dawley rats (Allred et al, 2004) *Genistein at 750 p.p.m. increased the weight of estrogen-dependent adenocarcinomas in ovariectomized rats compared with the negative-control animals.* Genistein also *enhanced mammary gland growth and tumor development of estrogen-dependent cells when ovariectomized mice were treated with the chemical carcinogen 1-methyl-1-nitrosourea to induce mammary tumorigenesis. Genistein treatment also resulted in a higher percentage of proliferative cells in tumors and increased uterine weights when compared with negative-control animals. In an endogenous estrogen environment similar to that of a postmenopausal woman, dietary genistein can stimulate the growth of a mammary carcinogen MNU-induced estrogen-dependent mammary tumors.* (Allred et al, 2004)

29. Use of multivitamins and prostate cancer mortality in a large cohort of US men. (Stevens et al, 2005) (#475,726 men who were cancer-free) *Regular multivitamin use was associated with a small increase in prostate cancer death rates.*

30 A randomized trial of antioxidant vitamins to prevent second primary cancers in head and neck cancer patients (Bairati et al, 2005 Apr 6) (#540 patients with stage I or II head and neck cancer treated by radiation therapy) Compared with patients receiving placebo, *patients receiving alpha-tocopherol supplements had a higher rate of second primary cancers during the supplementation period* but a lower rate after supplementation was discontinued. Similarly, the rate of having a recurrence or second primary cancer was higher during but lower after supplementation with alpha-tocopherol. CONCLUSIONS: *alpha-Tocopherol supplementation produced unexpected adverse effects on the occurrence of second primary cancers and on cancer-free survival.* Note: *Patients taking an antioxidant were 1.65 times more likely to suffer a return of their original cancer during the three years they were on the supplement. The risk was highest among those taking only vitamin E (1.86 times higher).*

31. Randomized trial of antioxidant vitamins to prevent acute adverse effects of radiation therapy in head and neck cancer patients (Bairati et al, 2005 Aug 20) (#540 patients with stage I or II head and neck cancer treated by radiation therapy) During the course of the trial, *supplementation with beta-carotene was discontinued because of ethical concerns.* Quality of life was not improved by the supplementation. *The rate of local recurrence of the head and neck tumor tended to be higher in the supplement arm of the trial.* Note: Researchers were concerned to find that *the rate of local recurrence (that is, a return of the original cancer) was 54 percent higher among patients on the combination pill than those on placebo.* There was a

smaller but still worrisome increase among those on vitamin E only. *This trial suggests that use of high doses of antioxidants as adjuvant therapy might compromise radiation treatment efficacy.*

32. Low-dose dietary phytoestrogen abrogates tamoxifen-associated mammary tumor prevention. (Lui et al, 2005) *In vitro studies of human and mouse **mammary tumor** cell lines confirm that low doses of genistein, co-administered with tamoxifen, promote cell proliferation.* In summary, *low-dose dietary isoflavones abrogated **tamoxifen-associated mammary tumor prevention** in vivo. Genistein attenuates the anti-tumor activities of tamoxifen.* (Lui et al, 2005)

33. Smoking, alcohol drinking, green tea consumption and the risk of esophageal cancer in Japanese men. (Ishikawa et al, 2006) (#9,008 men in Cohort 1 and 17,715 men in Cohort 2) *Cigarette smoking, alcohol drinking and green tea consumption were significantly associated with an increased risk of esophageal cancer. The population attributable fractions of esophageal cancer incidence that was attributable to smoking, alcohol drinking and green tea consumption were 72.0%, 48.6%, and 22.1%, respectively.*
CONCLUSIONS: Among the variables studied, *smoking has the largest public health impact on esophageal cancer incidence in Japanese men, followed by alcohol drinking and green tea drinking.*

34. Antioxidant Supplementation Increases the Risk of Skin Cancers in Women but Not in Men. (Hercberg et al, 2007) (#French adults, 7,876 women and 5,141 men. Total # = 13,017) *In women, the incidence of SC was higher in the antioxidant group* [adjusted hazard ratio (adjusted HR) = 1.68; P = 0.03]. **Conversely, in men, incidence did not differ between the 2 treatment groups.** Despite the small number of events, *the incidence of melanoma was also higher in the antioxidant group for women.*

35. National Institutes of Health State-of-the-Science Conference Statement: Multivitamin/Mineral Supplements and Chronic Disease Prevention (NIH State-of-the Science Panel. 2007). **reports from RCTs that noted excess lung cancer occurring in asbestos workers and smokers consuming β-carotene.** In addition, *esophageal cancer excess was found with long-term follow-up of older Chinese patients (the Linxian study by Blot et al.) treated with selenium, β-carotene, and vitamin E supplements* (Blot et al, 1993) (NIH State-of-the Science Panel. 2007)

36. Associations of antioxidant nutrients and oxidative DNA damage in healthy African-American and White adults (Watters JL, et al, 2007) **(#164)** *There was also a positive association of vitamin E with oxidative DNA damage in the total population and in African-American men after adjusting for covariates.* (Watters JL, et al, 2007)

37. Multivitamin use and risk of prostate cancer in the National Institutes of Health-AARP Diet and Health Study (Lawson et al, 2007) (#295,344 men) *Investigators found an increased risk of advanced and fatal prostate cancers among men reporting excessive use of multivitamins (more than seven times per week) when compared with never users. The* *positive associations with excessive multivitamin use were strongest in men with a family history of prostate cancer or who took individual micronutrient supplements, including selenium, beta-carotene, or zinc.* (Lawson et al, 2007)

38. Health Professionals Follow-up Study (2007): Effect of vitamins C, E, A and carotenoids and the occurrence of oral pre-malignant lesions (Maserejian et al, 2007) **(#42,340 men enrolled in the Health Professionals Follow-up Study) (#207 found with oral premalignant lesions)** *A trend for increased risk of oral pre-malignant lesions was observed with vitamin E, especially among current smokers and with vitamin E supplements. Beta-carotene also increased the risk among current smokers.* However, **dietary vitamin C was significantly associated with a reduced risk of oral premalignant lesions.**

39. VITAL (VITamins and Lifestyle) study (2008) (Slatore et al, 2008) (#77,721 men and women); *Supplemental vitamin E was associated with a small increased risk of lung cancer.*

40. Efficacy of Antioxidant Supplementation in Reducing Primary Cancer Incidence and Mortality: Systematic Review and Meta-analysis (Bardia et al, 2008) (#104,196 participants) *Beta carotene supplementation was associated with an increase in the incidence of cancer among smokers and with a trend toward increased cancer mortality.*

41. Multivitamin-multimineral supplement use and mammographic breast density. (Berube et al, 2008) (#Premenopausal (777) and postmenopausal (783) women; total 1,560) *Regular use of multivitamin-multimineral supplements may be associated with higher mean breast density among premenopausal women.*

42. Continuous in vitro exposure to low-dose genistein induces genomic instability in breast epithelial cells. (Kim et al, 2008) *Genistein exposure of L5178Y mouse lymphoma cells at concentrations less than 100 nM induced micronuclei formation and mutagenesis at the thymidine kinase locus* (Boos, G., Stopper, H., Genotoxicity of several clinically used topoisomerase II inhibitors. Toxicol. Lett. 2000, 116, 7–16) *and chronic exposure (3 months) of human MCF-10A cells induced genomic instability.* (Kim et al, 2008)

43. Dietary genistein negates the inhibitory effect of letrozole on the growth of aromatase-expressing estrogen-dependent human breast cancer cells. (Ju et al, 2008) *Dietary GEN increased the growth of MCF-7Ca tumors implanted in ovariectomized mice and could also negate the inhibitory effect of LET on MCF-7Ca tumor growth. GEN can reverse the inhibitory effect of LET on tumor growth and adversely impact breast cancer therapy.* Overall, **the data suggest that short-term exposure to soy caused a weak estrogenic effect.** *An additional concern about the potential harm that might be done by genistein stems from data that it can stimulate proliferation of tamoxifen-sensitive cells.* (Ju et al, 2008)

44. Phytoestrogens and breast cancer: a complex story. (Helferich et al, 2008) *Dietary genistein was able to negate the inhibitory effect of TAM on E-stimulated tumor growth.* In summary, *genistein can act as an estrogen agonist resulting in proliferation of E-dependent* *human **breast cancer** tumors in vivo*. Additionally, dietary genistein can negate the inhibitory effects of TAM on E-stimulated growth of MCF-7 cell tumors implanted into ovariectomized athymic mice. (Helferich et al, 2008)

45. Effect of selenium and vitamin E on risk of prostate cancer and other cancers: The Selenium and Vitamin E Cancer Prevention Trial (SELECT) (2009) (Lippman et al, 2009, SELECT) (#35,533 men) **There were statistically nonsignificant increased risks of prostate cancer in the vitamin E group** but not in the selenium + vitamin E group. CONCLUSION: **Selenium or vitamin E, alone or in combination at the doses and formulations used, did not prevent prostate cancer in this population of relatively healthy men.** *The trial was stopped ahead of its original 12-year deadline because of a lack of any noticeable benefit.*

46. Total and Cancer Mortality After Supplementation with Vitamins and Minerals: 10-year Follow-up of the Linxian General Population Nutrition Intervention Trial. (Qiao et al, 2009) (#29,584 adult participants) *esophageal cancer deaths increased 14% among those aged 55 years or older. Vitamin A and zinc supplementation was associated with increased total and stroke mortality.*

47. Long-term use of beta-carotene, retinol, lycopene, and lutein supplements and lung cancer risk: results from the VITamins and Lifestyle (VITAL) study. (Satia et al, 2009) (#77,126 (VITAL) *Longer duration of use of individual beta-carotene, retinol, and lutein supplements (but not total 10-year average dose) was associated with statistically significantly elevated risk of total lung cancer and histologic cell types.*

48. Vitamin and mineral use and risk of prostate cancer: the case-control surveillance study. (Zhang et al. 2009) **(#1,706 prostate cancer cases and 2,404 matched controls).** *Men who used zinc for ten years or more, either in a multivitamin or as a supplement, had an approximately two-fold increased risk of prostate cancer. The finding that long-term zinc intake from multivitamins or single supplements was associated with a doubling in risk of prostate cancer adds to the growing evidence for an unfavorable effect of zinc on prostate cancer carcinogenesis.*

49. Folic acid and risk of prostate cancer: results from a randomized clinical trial. (Figueiredo et al, 2009) **(643 randomly assigned men).** *the estimated probability of being diagnosed with prostate cancer over a 10-year period was 9.7% in the folic acid group and 3.3% in the placebo group.*

50. Green tea consumption and risk of stomach cancer: a meta-analysis of epidemiologic studies (Myung, Int J Cancer. et al, 2009) **(#13 epidemiologic studies)** *In the meta-analyses of the recent cohort studies, the highest green tea consumption was shown to significantly increase stomach cancer risk using the crude data,* but no significant association between them was seen when using the adjusted data.

51. Green tea (Camellia sinensis) for the prevention of cancer. Cochrane Database Syst Rev. 2009 Jul 8;(3):CD005004. (Boehm et al, 2009) (#Fifty-one studies with more than 1.6

million participants were included) **there was limited to moderate evidence that the consumption of green tea reduced the risk of lung cancer, especially in men, and** *urinary bladder cancer or that it could even increase the risk of the latter*.

52. Molecular consequences of genetic variations in the glutathione peroxidase 1 selenoenzyme. (Zhou et al, 2009) *Polymorphisms in the human GPx-1 gene are associated with increased risk of cancer.* (Zhou et al, 2009)

53. In vitro evaluation of selenium genotoxic, cytotoxic, and protective effects: **a review** (Valdiglesias et al, 2009) There *exists a considerable literature indicating both cytotoxic and genotoxic effects of selenium when cells in culture or animals are provided high doses, and consequences include enhanced mutagenesis, induction of chromosomal abnormalities such as micronuclei formation, as well as induction of cell-cycle arrest and apoptosis.* (Valdiglesias et al, 2009)

54. Developmental toxicity and brain aromatase induction by high genistein concentrations in zebrafish embryos. *High concentrations of genistein caused a teratogenic effect on zebrafish embryos and confirmed the estrogenic potential of genistein in mosaic reporter zebrafish embryos. Additional concerns about possible teratogenic effects of genistein have been raised in the studies using zebrafish.* (Kim et al, 2009) (Sassi-Messai et al, 2009)

55. Antioxidant supplement use is associated with melanoma risk in light of recently published data from the Supplementation in Vitamins and Mineral Antioxidants (SUVIMAX) study, which reported a 4-fold higher melanoma risk in women. (Asgari et al, 2009) referred to data from the SUVIMAX study and not Asgari's study.

56. Multivitamin use and breast cancer incidence in a prospective cohort of Swedish women. (Larsson et al, 2010) (#35,329 cancer-free women) *Multivitamin use was associated with a statistically significant increased risk of breast cancer. Use of multivitamins was linked to a statistically significant 19 per cent increased risk of breast cancer* (after adjusting for lifestyle and risk factors like weight, diet, smoking, exercise, and family history of breast cancer.

57. Quercetin and Ferulic Acid Aggravate Renal Carcinoma in Long-Term Diabetic Victims. (Chiu-Lan Hsieh et al, 2010). Conclusively, the phytoantioxidants *quercetin and ferulic acid are able to aggravate, if not induce, nephrocarcinoma in mice.*

58. Selenomethionine and alpha-tocopherol do not inhibit prostate carcinogenesis in the testosterone plus estradiol-treated NBL rat model (Ozten et al, 2010) *Alpha-Tocopherol significantly increased the incidence of adenocarcinomas of the mammary glands at both dietary concentrations in rats.* **Importantly, the results of the current animal studies and those reported previously were fully predictive of the outcome of the Selenium and Vitamin E Cancer Prevention Trial. (Ozten et al, 2010)**

59. Semen quality and sperm DNA damage in relation to urinary bisphenol A among men from an infertility clinic. (Meeker et al, 2010) **Semen quality and sperm DNA damage in**

relation to urinary bisphenol A among men from an infertility clinic (Meeker et al, 2010) (#190) *Urinary* Bisphenol A (BPA) (an antioxidant) *may be associated with declined semen quality and increased sperm DNA damage.* (Meeker et al, 2010) not numbered in 150 list

60. Incidence of skin cancers during 5-year follow-up after stopping antioxidant vitamins and mineral supplementation (Ezzedine et al, 2010) (#12,741) In the SU.VI.MAX study, antioxidant supplementation for 7.5 years was found to increase skin cancer risk in women but not in men. **randomised trial,** (daily a placebo or a combination of ascorbic acid (120 mg), vitamin E (30 mg), β-carotene (6 mg), selenium (100 µg) and zinc (20mg), from inclusion in 1994 to September 2002) *The risk of skin cancers associated with antioxidant intake declines following interruption of supplementation. This supports a causative role for antioxidants in the evolution of skin cancers.* (Ezzedine et al, 2010)

61. Source-specific effects of micronutrients in lung cancer prevention (Roswall et al, 2010, Lung cancer) (#55,557) **vitamin C, E, folate and beta-carotene and lung cancer** risk. **We found a significant protective effect of dietary vitamin E intake** *and a significantly higher lung cancer risk with supplemental beta-carotene and dietary folate intake.* (Roswall et al, 2010, Lung cancer)

62. The protective effects of nutritional antioxidant therapy on Ehrlich solid tumor-bearing mice depend on the type of antioxidant therapy chosen: histology, genotoxicity and hematology evaluations. (Miranda-Vilela AL, et al, 2011) **Antioxidant administrations before tumor inoculation effectively inhibited its growth in the three experimental protocols,** but *administrations after the tumor's appearance accelerated tumor growth and favored metastases.* **Continuous administration of pequi oil inhibited the tumor's growth, while the same protocol with** *vitamins E and C accelerated it (tumor growth), favoring metastasis and increasing oxidative stress on erythrocytes.*

63. Prenatal exposure to flavonoids: Implication for cancer risk (Vanhees et al, 2011) *In vitro exposure to genistein/quercetin induced higher numbers of Mll rearrangements in bone marrow cells of Atm-ΔSRI mutant mice compared with wt mice. Prenatal exposure to flavonoids associated with higher frequencies of Mll rearrangements and a slight increase in the incidence of malignancies in DNA repair-deficient mice.* These data suggest that *prenatal exposure to both genistein and quercetin supplements could increase the risk on Mll rearrangements especially in the presence of compromised DNA repair.*

64. Use of vitamin supplements and risk of total cancer and cardiovascular disease among the Japanese general population: a population-based survey (Hara et al, 2011) (#28,903 men and 33,726 women for a total of 62,629) In women, consistent vitamin supplement use was associated with lower risk of CVD, *whereas past and recent use were associated with higher risk of cancer.* (Hara et al, 2011)

65. Antioxidants tied to mixed effects in breast cancer. (Greenlee et al, 2011) *Women who regularly took a mix of carotenoids had a higher risk of dying from breast*

cancer. Dietary supplements containing high doses carotenoids may be harmful, and people should think twice before taking them." (Greenlee et al, Cancer 2011)

66. Vitamin E pills linked with prostate cancer risk. (Klein et al, JAMA, 2011) Men randomly assigned to take a 400-unit capsule of vitamin E every day for about five years were 17 percent more likely to get prostate cancer than those given dummy pills.

67. Potential utility of natural products as regulators of breast cancer-associated aromatase promoters. (Khan SI, et al, 2011) *Genistein was shown to increase aromatase activity in human adrenocortical carcinoma (H295R) cells and in isolated rat ovarian follicles (thus, encouraging their growth).* (Khan SI, et al, 2011)

68. Fish oils can block chemotherapy drugs (Prof Emile Voest, 2011) Fats found in fish oil supplements can stop chemotherapy drugs working in mice. (Voest et al, 2011) Researchers showed that *off-the-shelf fish oil supplements, given to mice, could stop chemotherapy working against some tumors.* These fatty acids were "abundantly present in commercially available fish oil products." *They currently recommend that these products should not be used whilst people are undergoing chemotherapy* (Paul Emile Voest, Univ Med Ctr Utrecht, Cancer Cell 2011)

69. The combined influence of multiple sex and growth hormones on risk of postmenopausal breast cancer: a nested case-control study (Tworoger et al, 2011) (#265 cases and 541 controls in the prospective Nurses' Health Study) *Women in the top versus bottom quintile of individual estrogen or androgen levels had approximately a doubling of postmenopausal breast cancer risk (both are antioxidants). Multiple hormones with high circulating levels substantially increase the risk of breast cancer, particularly ER-positive disease.* **Elevated estrogens had the biggest effect on cancer risk, especially for ER-positive cancer.** (Tworoger et al, 2011)

70. Selenium and vitamin E for prostate cancer: post-SELECT (Selenium and Vitamin E Cancer Prevention Trial) status (Ledesma et al, 2011) *SELECT was terminated early because of both safety concerns and negative data for the formulations and doses given.* (Ledesma et al, 2011)

71. The role of antioxidant supplement in immune system, neoplastic, and neurodegenerative disorders: a point of view for an assessment of the risk/benefit profile (Brambilla et al, 2008)

Although evidence shows that antioxidant treatment results in cytoprotection, the potential clinical benefit deriving from both nutritional and supplemental antioxidants is still under wide debate. In this line, *the inappropriate assumption of some lipophilic vitamins has been associated with increased incidence of cancer rather than with beneficial effects.* (Brambilla et al, 2008)

72. Safety of vitamins and minerals: controversies and perspective (Soni et al. 2010)

Numerous studies published over more than a decade have linked some supplements (including vitamins E, C, D, A, and B, as well as selenium) to no health benefits or even to adverse health effects. *Recent studies with negative results, which drew media attention, include the following: a 2008 study on the ability of vitamin E and selenium to lower the risk of prostate cancer was halted amidst fear of potential harm; vitamin C may do more harm than good as it may protect cancer cells;* intake of vitamins E and C by 15,000 male physicians for 10 years had no health benefits. (Soni et al. 2010)

HEART DISEASE (28 Study Reports)

1. α-Tocopherol, β-Carotene Cancer Prevention Study (ATBC study) (Heinonen et al, 1994) (#29,133 men); *50% increase in hemorrhagic stroke deaths among vitamin E group; 11% increase in ischemic heart disease deaths among β-carotene group; 18% increase in lung cancer among β-carotene group.*

2. The β-Carotene and Retinol Efficacy Trial (CARET) (Omenn et al, 1996) (#14,254 heavy smokers and 4,060 asbestos workers) (total #18,314 men and women); *28% increase in lung cancer; 26% increase in CVD (nonsignificant); 17% increase in total mortality* among *treatment group. This study was stopped 21 months earlier than planned.*

3. ATBC Sub-Study Shows Increased CVD Deaths (Rapola et al, 1997) (#1,862 men, with prior myocardial infarction); there **were no significant differences** in major coronary events but *significantly more deaths from fatal coronary heart disease.* **There were no significant differences in the number of major coronary events between any supplementation group and the placebo group. There were** *significantly more deaths from fatal coronary heart disease in the beta-carotene and combined alpha-tocopherol and beta-carotene groups* **than in the placebo group.** *The risk of fatal coronary heart disease increased in the groups that received either beta-carotene or the combination of alpha-tocopherol and beta-carotene.* **They do not recommend the use of alpha-tocopherol or beta-carotene supplements in this group of patients.**

4. A report from the **Atherosclerosis Risk In Communities (ARIC)** group found individuals with *the highest carotid IMT to have lower levels of plasma carotenoids but higher alpha-tocopherol and retinol levels compared to controls* (Iribarren et al, 1997). This study is not listed in 430 reports.

5. The Multivitamins and Probucol Study (Tardif et al, 1997) (#317 participants); *Probucol has been pulled off the market due harmful effects and the likelihood of cardiac arrhythmias.*

6. The influence of antioxidant nutrients on platelet function in healthy volunteers (Calzada et al, 1997) (#40 healthy volunteers) **Supplementation of healthy volunteers with** *vitamin E* *decreased platelet function* whereas supplementation with vitamin c or beta-carotene had no significant effects.

7. The Multivitamins and Probucol Study (Tardif et al, 1997) (#317 participants) Probucol has been pulled off the market due harmful effects and the likelihood of cardiac arrhythmias. (Tardif et al, 1997)

8. Familial hypercholesterolemia, intima-to-media thickness (FH IMT study) (Raal et al, 1999) (#15 with homozygous familial hypercholesterolemia); *homozygous familial hypercholesterolemia, intima-to-media thickness (FH IMT study) increased with vitamin E supplements* (400 mg/day) for 2 years.

9. Vitamin E Worsens Metabolic Parameters in Type 2 Diabetics. (Skrha et al. 1997) (#12) *decreases of glucose disposal rate, metabolic clearance rate of glucose, and insulin receptor number were found after vitamin E administration as compared with pretreated values. A worsening of diabetes control as observed by an increase of HbA1C was present.*

10. HDL Atherosclerosis Treatment study (HATS) (Brown et al, 2001) (#160 participants); an antioxidant cocktail (vitamin E, ß-carotene, vitamin C, and selenium) had a 0.7% progression in stenosis after 3 years, compared with 0.4% regression in the group on only simvastatin/niacin. Thus, *antioxidant supplements may have interfered with the efficacy of statin-plus-niacin therapy.* No clinical or angiographically measurable benefit from antioxidants was found. *When used in combination with simvastatin/niacin, antioxidants negated the benefit of the latter on plasma lipid profile and stenosis progression.*

11. Women's Angiographic Vitamin and Estrogen (WAVE) Trial (Waters et al, 2002) (#423 postmenopausal women, with at least one 15% to 75% coronary stenosis); neither HRT nor antioxidant vitamin supplements (vitamins C & E) provided any cardiovascular benefit. Instead, *a potential for harm was suggested. There is some **evidence of potentially adverse effects of antioxidant supplements on CVD as assessed by angiographic end points**. In the Women's Angiographic Vitamin and Estrogen Study, **postmenopausal women with coronary disease on hormone replacement therapy given vitamin E plus vitamin C had an unexpected significantly higher all-cause mortality rate and a trend for an increased cardiovascular mortality rate** compared with the vitamin placebo women.*

12. Plasma carotenoids and tocopherols and risk of myocardial infarction in a low-risk population of US male physicians. (Hak et al. 2003) (#531 physicians diagnosed with MI) that *men with high plasma gamma-tocopherol levels tended to have an increased risk of nonfatal and fatal MI.*

13. Comorbidities in gouty arthritis (Marwah, 2011) Gouty arthritis is increasing in prevalence in men and women, particularly in older age groups. *Gouty arthritis is associated with numerous comorbidities that are increasing in prevalence (chronic kidney disease [CKD], hypertension, obesity, diabetes, metabolic syndrome, and cardiovascular disease) and that negatively impact long-term prognosis and quality of life.* Gouty arthritis is associated with renal, metabolic, and cardiovascular comorbidities that negatively impact overall health.

(Marwah, 2011) **RMH Note: Uric acid is the most prevalent antioxidant in the body and excessive levels are especially harmful.**

14. Supplemental vitamin C increase cardiovascular disease risk in women with diabetes (Iowa Women's Health study) (Lee et al, 2004) (#1,923 postmenopausal women who reported being diabetic) *A high vitamin C intake from supplements is associated with an increased risk of cardiovascular disease mortality in postmenopausal women with diabetes.*

15. ATBC 6-year followup study (2004) (Thornwall et al., 2004) (#29,133 male smokers); *β-Carotene seemed to increase the post-trial risk of first-ever non-fatal MI.*

16. Vitamin C worsens coronary atherosclerosis in those with two copies of the haptoglobin 2 gene. (Levy et al, 2004) (#299 postmenopausal women) *antioxidant therapy (1,000 mg/day of vitamin C + 800 IU/day of vitamin E) was associated with improvement of coronary atherosclerosis in diabetic women with two copies of the haptoglobin 1 gene but worsening of coronary atherosclerosis in those with two copies of the haptoglobin 2 gene.*

17. HOPE-TOO Extension (Lonn et al, 2005) (#3,994 original study enrollees) *Another subgroup finding in HOPE-TOO was a vitamin E–associated increased risk of heart failure incidence that appeared in a secondary end point analysis in the 4.5-year report and persisted in the 7-year extended follow-up, as did the risk of hospitalization for heart failure. Patients in the vitamin E group had a higher risk of heart failure and hospitalization for heart failure.*

18. Fish oil supplementation and risk of ventricular tachycardia and ventricular fibrillation in patients with implantable defibrillators (Riatt et al, 2005) (#200) *Among patients with a recent episode of sustained ventricular arrhythmia and an ICD, fish oil supplementation does not reduce the risk of VT/VF and may be proarrhythmic in some patients.* (Riatt et al, 2005) (Riatt et al, Fish oil supplementation and risk of ventricular tachycardia and ventricular fibrillation in patients with implantable defibrillators. JAMA. 2005;293(23):2884-2891)

19. The rate of low-birth-weight babies was higher and the rate for gestational hypertension was higher for women in the vitamin group. Women in the vitamin group had an increased risk of being hospitalized antenatally for hypertension and having to take antihypertensive medication. In addition, **a subgroup of women in the vitamin group had a higher frequency of abnormal liver-function tests.** (Rumbold et al, 2006)

20. The Melbourne Atherosclerosis Vitamin E Trial (MAVET): a study of high dose vitamin E in smokers. (Magliano et al, 2006) (#409 male and female smokers) *The mean increase in intima-media thickness over time in the vitamin E group was 0.0041 mm/year faster than placebo.*

21. The effect of vitamin E on blood pressure in individuals with type 2 diabetes: a randomized, double-blind, placebo-controlled trial (Ward et al, 2007) (#58 with type 2 diabetes randomized) *Treatment with alpha-tocopherol significantly increased systolic BP, diastolic BP, pulse pressure and heart rate versus placebo. Treatment with mixed tocopherols* *significantly increased systolic BP, diastolic BP, pulse pressure and heart rate versus placebo. In contrast to our initial hypothesis, treatment with either alpha- or mixed tocopherols significantly increased BP, pulse pressure and heart rate in individuals with type 2 diabetes.* (Ward, Wu et al, 2007)

22. *Treatment with the combination of vitamin C and polyphenols increased systolic BP and diastolic BP. These results suggest caution for hypertensive subjects taking supplements containing combinations of vitamin C and polyphenols.* (Ward, Hodgson et al, 2007)

23. *Treatment with the combination of vitamin C and polyphenols increased systolic BP and diastolic BP. These results suggest caution for hypertensive subjects taking supplements containing combinations of vitamin C and polyphenols.* (Ward, Hodgson et al, 2005)

24. N-acetylcysteine to reduce renal failure after cardiac surgery: a systematic review and meta-analysis (Naughton et al. 2008) (#Seven randomized controlled trials (RCTs, n = 1000) There was a small, though significant **increase in postoperative blood loss after cardiac surgery among patients treated with NAC.** (Naughton et al, 2008)

25.Heterogeneity in randomized controlled trials of long chain (fish) omega-3 fatty acids in restenosis, secondary prevention and ventricular arrhythmias (Jenkins et al, 2008)

Randomized controlled trials of marine omega-3 fatty acid supplementation in relation to coronary heart disease (CHD) have inconsistent outcomes, yet public health messages are uniformly positive. *Newer data indicate that fish oils may increase CHD events in men with angina.* Furthermore, *in two of three trials in patients with implantable cardioverter defibrillators (ICDs), and a history of ventricular arrhythmias, fish oils showed no significant benefit or even increased the risk of appropriate ICD discharge.* Due to significant heterogeneity in the response to fish oils, further studies are required before making widespread recommendations for all groups to increase consumption of fish and fish oil. (Jenkins et al, 2008)

26. Effects of long-term antioxidant supplementation and association of serum antioxidant concentrations with risk of metabolic syndrome in adults (SU.VI.MAX) (Czernichow et al, 2009) **(#5,220 adults) Adults (*n* = 5,220) Antioxidant supplementation for 7.5 y did not affect the risk of metabolic syndrome (MetS); *Baseline serum antioxidant concentrations of β-carotene and vitamin C, however, were negatively associated with the risk of MetS.***

27. Oral antioxidants and cardiovascular health in the exercise-trained and untrained elderly: a radically different outcome (Wray, 2009) (#6) *Antioxidant administration after exercise training negated these improvements (reduced BP and increased FMD), returning subjects to a hypertensive state and blunting training-induced improvements in FMD (flow-mediated vasodilation).* In conclusion, the paradoxical effects of these interventions suggest a need for caution when exercise and acute antioxidant supplementation are combined in elderly mildly hypertensive individuals. (Wray et al, 2009)

28. Micronutrient concentrations and subclinical atherosclerosis in adults with HIV. (Falcone et al, 2010) (#298 Nutrition for Healthy Living participants) *elevated serum vitamin E concentrations are associated with abnormal markers of atherosclerosis and may increase the risk of cardiovascular complications in HIV-infected adults*.

STROKE (6 Study Reports)

1. α-Tocopherol, β-Carotene Cancer Prevention Study (ATBC study) (Heinonen et al, 1994) (#29,133 men); *50% increase in hemorrhagic stroke deaths among vitamin E group; 11% increase in ischemic heart disease deaths among β-carotene group; 18% increase in lung cancer among β-carotene group.*

2. Controlled trial of alpha-tocopherol and beta-carotene supplements on stroke incidence and mortality in male smokers. (Leppala et al. 2000) (#28,519 male cigarette smokers). *alpha-Tocopherol supplementation increased the risk of subarachnoid hemorrhage 50% but decreased that of cerebral infarction 14%, whereas beta-carotene supplementation increased the risk of intracerebral hemorrhage 62%. alpha-Tocopherol supplementation also increased the risk of fatal subarachnoid hemorrhage 181%.* The overall net effects of either supplementation on the incidence and mortality from total stroke were nonsignificant. *alpha-Tocopherol supplementation increases the risk of fatal hemorrhagic strokes* but prevents cerebral infarction. The effects may be due to the antiplatelet actions of alpha-tocopherol. *beta-Carotene supplementation increases the risk of intracerebral hemorrhage.*

3. Tolerability of NXY-059 at higher target concentrations in patients with acute stroke (Lees, 2003) (#134) In patients with acute stroke, NXY-059, a nitrone-based free radical-trapping agent (antioxidant), *serious adverse events occurred in 3, 17, and 13 patients, respectively, with deaths in 0, 4, and 3 patients and treatment discontinuations because of adverse events in 0, 1, and 3 patients.* Good outcome, defined by modified Rankin Scale score of 0 or 1, was seen in 53%, 29% and 40%, respectively. (Lees, 2003)

4. Vitamins E and C in the prevention of cardiovascular disease in men: The Physicians' Health Study II randomized controlled trial (Sesso et al, 2008) (#14,641 US male physicians) *vitamin E was associated with an increased risk of hemorrhagic stroke.*

5. Total and Cancer Mortality After Supplementation with Vitamins and Minerals: 10-year Follow-up of the Linxian General Population Nutrition Intervention Trial. (Qiao et al, 2009) (#29,584 adult participants) *esophageal cancer deaths increased 14% among those aged 55 years or older. Vitamin A and zinc supplementation was associated with increased total and stroke mortality.*

6. Effects of vitamin on stroke subtypes: meta-analysis of randomized controlled trials. (Schurks et al. 2010) (#118,765) In this meta-analysis, *vitamin E increased the risk for hemorrhagic stroke by 22% and reduced the risk of ischemic stroke by 10%.*

Human Cancer Cell Types (27 Human & 9 Murine) Shielded by Antioxidants

If you are worried about cancer, just take a look at the human cell types that are protected by antioxidants from cell death in lab experiments.

Unbelievably, there are twenty-seven (27) types of human cancer cell types and nine (9) murine cancer cell types that can be killed by EMODs and in which the killing can be blocked by antioxidants, thereby providing antioxidant protection and shielding of the cancer cells. Published data has shown that antioxidants blocked the killing of the following human and murine (rodent) cancer cell types by EMODs:

- **human breast cancer** (J. Nutr. 134, 2004) (Gundimeda et al, 1996) (Peralta et al, 2006) (Aykin-Burns et al, 2009) (Xiao et al, Mol Cancer Ther. 2006)
- **human prostate carcinoma** (Xiao et al, 2006) (Wu et al, 2005) (Singh et al, 2005) (Cho et al, 2005) (Milanesa et al, 2000)
- **human non-small cell lung cancer** (Ling et al, 2003) (Wu et al, 2006)
- **human colon adenocarcinoma** (Wenzel et al, 2005)
- **human colon cancer** (Wenzel et al, 2004) (Aykin-Burns et al, 2009)
- **human colorectal carcinoma** (Chen et al, 2004) (Gali-Muhtasib et al, 2008)
- **human ovarian cancer cells** (Pak et al, 2011)
- **human melanoma** (Marcin et al, 2005) (Okroj et al, 2006) (Nishikawa et al, 2004) (Grimm et al, 2011)
- **human metastatic melanoma** (Kirshner et al, 2008)
- **human head and neck cancer** (Mattson et al, 2009) (Simons et al, 2007)
- **human lymphoma** (J. Nutr. 134, 2004) (Mansat-De Mas et al, 1999)
- **human leukemia** (Hileman et al, 2004) (McKallip et al, 2006) (Hou et al, 2005) (Feng et al, 2007) (Yedjou et al, 2008) (Hiraoka et al, 1998)
- **human hepatoma** (Wu et al, 2004) (Wu, Ng, Lin, 2004)
- **human hepatocellular liver carcinoma** (Shimoda et al, 2003)
- **human pancreatic cancer** (Maehara et al, 2004)
- **human multiple myeloma** (Grad et al, 2001) (Ahmad et al, 1997) (Gupta et al, 2000) (Nakazato et al, 2005) (Isham et al, 2007)
- **Burkitt's lymphoma** (Ahmad et al, 1997) (Gupta et al, 2000) (Nakazato et al, 2005) (Ahmad et al, 1997)
- **human chronic lymphocytic leukemia** (Kay, 2006) (Chandra et al, 2003) (Shanafelt et al, 2005) (Mow et al, 2002) (Biswas S, et al, 2010)
- **human acute myeloid leukemia** (Kay, 2006) (Chandra et al, 2003) (Shanafelt et al, 2005) (Mow et al, 2002)
- **human promyelocytic leukemia** (Hou et al, 2005)
- **human erythromyeloid leukemia** (Wagner et al, 2000)
- **human epithelial cancer cells** (breast and colon) (Aykin-Burns et al, 2009)
- **human endometrial cancer** (Llobet et al, 2008)
- **human bladder cancer cells** (Miyajima et al, 1999)
- **human invasive bladder cancer** (Miyajima et al, 1999 - human bladder cancer KU-1 cell line)
- **human glioblastoma cells** (Lee et al, 2004)

- **human osteosarcoma** (Ahmad et al, 2005)
- **murine pheochromocytoma** (Jang, Surh, 2001)
- **murine retinoblastoma** (Salganik et al, 2000)
- **murine thymoma** (Tome et al, 2001)
- **murine lymphoma- six cell types** (Nathan et al, vol 153, 1981)
- **murine leukemia** (Wagner et al, 1996)
- **murine fibrosarcoma** (Teicher et al, 1994)
- **murine neuroblastoma** (Prasad et al, PNAS. 1979)
- **murine mammary cancer** (Bracke et al, 1999)
- **murine brain cancer** (Zeisel (2), 2004)

There are twenty-seven (27) types of human cancer cell types and 9 murine cancers that can be killed by EMODs and in which the killing can be blocked by antioxidants, thereby providing antioxidant protection and shielding of the cancer cells. Part of this list was compiled in 2009. (Howes, Philica. Feb 7, 2009).

Part of this list was compiled in 2009 and has been updated for this book. (Howes, Philica. Feb 7, 2009). **It is evident to many investigators that the in vitro apoptogenic agents function as prooxidants.** (Hail et al, 2008). Note: References for this list are available in my book, *Dangers of Excessive Antioxidants in Cancer Patients, 2011*.

Listing of Human Cancer Cell Lines Killed by EMOD Induced Apoptosis: This material was compiled specifically for this 2016 book.

Human lung carcinoma cells A549 killed by DADS via EMOD induced apoptosis, blocked by NAC (Wu, Kassie, Mersh-Sundermann, 2005). ==
Human lung carcinoma A549 cells, blocked by NAC (Wu et al, 2006). ==
Human melanoma A375 cells and basal cell carcinoma (BCC) cells killed by DATS via EMOD induced apoptosis, blocked by NAC (Wang et al, 2012). ==
Human HCT-116 colon cancer cells killed by DADS via EMOD induced apoptosis, blocked by NAC (Song et al, 2009). ==
Human breast cancer cells killed by garlic derived compounds (Tsubura et al, 2011). =
Pancreatic cancer cell line KLM1 killed by Gencitabine EMOD induced apoptosis, blocked by selenoprotein P (Maehara et al, 2004). ==
Human hepatoma cells HepG2 killed by pericarp extract via EMOD induced apoptosis, blocked by Trolox (Paik et al, 2005). ==
Human colon cancer cells HT-29 and HCT-116 killed by Annona muricata via EMOD induced apoptosis (Moghadamtousi, et al, 2014). =
Lung cancer cells, A549 killed by EMODs, blocked by SOD and CAT (Moghadamtousi, Kadir, Paydar et al, 2014). ==
Human lung adenocarcinoma cells A549 killed by chamaejasmine via EMOD induced apoptosis (Yu et al, 2011). =
Human adenocarcinoma alveolar basal epithelial A549 cells killed by Baohuoside 1 via EMOD induced apoptosis, blocked by NAC (Song et al, 2012). ==

Human A549 lung adenocarcinoma cells killed by curcumin via EMOD induced apoptosis, blocked by caspases inhibitor, Z-VAD-fmk (Chen, Lu, Wu et al, 2010). ==

Human MCF-7 breast cancer cell line, CAPAN-1 pancreatic cancer cell line, PC-3 prostate cancer cell lines (Z-138, Jeko-1, Granta) and SP53 Non-Hodgkin's lymphoma cell lines, JURKAT T-cell acute lymphoblastic leukemia cells, and MCF 10A immortalized mammary epithelial cell line, glioma (LN229 and U87MG) cell lines, were killed by iron ligand metallodrugs via EMOD induced apoptosis, and were blacked by NAC (Gonzalez-Bartulos et al, 2015). ==

Human HT29 and HCT116 colon cancer cells killed by EMODs (Fath et al, 2009) =

Paclitaxel (PTX) increases EMOD breast cancer cell kill, blocked by NAC (Hadzic et al, 2010). ==

Human leukemia HL-60 cells killed by vitamin K3 via EMOD induced apoptosis, blocked by CAT (Lin et al, 2005). ==

Human colon and breast cancer cells (HT29, HCT116, SW480 and MB231) and HMECs (human mammary epithelial cells) apoptosis, blocked by SOD and CAT (Aykin-Burns et al, 2009). ==

Tamoxifen kills breast cancer cells via EMOD induced apoptosis, blocked by vitamin E (Gundimeda et al, 1996). ==

Human multiple myeloma cell lines: 8226/S, 8226/Dox40, U266, and U266/Bcl-x(L)Multiple myeloma killed by As_2O_3 EMOD induced apoptosis, blocked by NAC (Grad et al, 2001). ==

Human head and neck cancer cells (FaDu) killed by cisplatin via EMOD induced apoptosis, blocked by NAC (Simons et al, 2007). ==

FaDu, Cal-27, and SQ20B head and neck cancer cells killed by EMODs, blocked by NAC (Mattson et al, 2009). ==

Human pancreatic cancer cells killed by EMODs and dicumarol: blocked by MnSOD (Cullen et al, 2003). ==

Human cancer cell lines (PC-3, DU145, MDA-MB231, and HT-29) and human osteosarcoma cells EMOD apoptosis blocked by SOD and CAT (Ahmed et al, 2005). ==

Resveratrol prooxidant activity killed human leukemia cells by EMODs induced apoptosis (Ahmad et al, 2003) =

Human leukemia cells PLB-985 killed by EMODs, blocked by CAT (Hiraoka et al, 1998). ==

Human prostate epithelial cells (P69SV40T) killed by SOD over expression EMOD induced apoptosis (Venkataraman et al, 2005). =

Prostate cancer cells PC-3 killed by hyperthermia EMOD induced apoptosis, blocked by MnSOD overexpression (Venkataraman et al, 2008). =

HL-60 promyelocytic leukemia cell line killed by peroxide induced apoptosis, blocked by myeloperoxidase inhibitor (Wagner et al, 2000). ==

Human prostatic cancer PC-3 cells killed by Methylglyoxal-induced apoptosis (Milanesa et al, 2000). =

Stage IV metastatic melanoma suppressed by Elesclomol (formerly STA-4783) oxidative EMOD apoptosis: blocked by NAC (Kirshner et al, 2008). ==

Human melanoma MDA-MB-435S cell lines killed by N-ß-Alanyl-5-S-glutathionyl-3,4 dihydroxyphenylalamine (5-S-GAD) via EMOD induced apoptosis, blocked by CAT (Nishikawa et al, 2004). ==

Human breast and renal cell carcinoma cells killed by Calcitriol via EMOD induced apoptosis, blocked by NAC and SOD (Ravid, Koren, 2003) =

Human P388 lymphoma cells killed in vivo by EMOD induced apoptosis (peroxide) (Nathan, Cohn, 1981). =

CLL/SLL Hodgkin lymphoma and non-Hodgkin lymphoma cells killed by PCI-24781 EMOD induced apoptosis (Bhalla et al, 2009). =

CD138⁺ myeloma cells killed in vivo by Chaetocin EMOD induced apoptosis, blocked by glutathione (Isham et al, 2007). ==

Human prostate cancer cells killed by GLIPR1 induced EMOD apoptosis (Li et al, 2008). =

Human A549 lung cancer cells killed by motexafin gadolinium (MGd) via EMOD induced apoptosis (Lecane et al, 2005). =

Human DU145 prostate cancer cells killed by sulforaphane EMODs, blocked by NAC (Cho et al, 2205). ==

Human PC-3 and DU145 prostate cancer cells killed by sulforaphane EMODs, blocked by NAC (Singh et al, 2005). ==

EMOD selective kill of ovarian cancer cells with beta-phenylethyl isothiocyanate (PEITC) (Wu, Hua, 2007). =

Human hepatoma PLC/PRF/5 (CD95-negative) cells Phenylethyl isothiocyanate (PEITC) apoptosis, blocked by NAC and Vitamin E (Wu et al, 2005). ==

U937 histiocytic lymphoma cells killed by ceramide EMOD induced apoptosis, blocked by NAC (Mansat-De Mas et al, 1999). ==

MCF-7 breast cancer cells killed by EMOD induced apoptosis (Farah and Begum, 2003). =

Ovarian cancer recurrence prevented by vitamin C EMOD induced apoptosis (Drisko et al, 2003). =

Human Burkitt's lymphoma cells (JLP-119) killed by vitamin C via EMOD induced apoptosis (Chen et al, 2005). =

Arsenic trioxide and ascorbic acid demonstrate promising activity against primary human chronic lymphocytic leukemia (CLL) Hu1D10 cells in vitro, blocked by NAC (Biswas et al, 2010). ==

MIA PaCa-2, AsPC-1, and BxPC-3 human pancreatic adenocarcinoma cells killed by Vitamin C via EMOD induced apoptosis, blocked by CAT (Du et al, 2010). ==

PS-341 (bortezomib, Velcade) kills relapsed human multiple myeloma via EMOD induced apoptosis, blocked by vitamin C (Zou et al, 2006). ==

Human HT-29 colon carcinoma cells killed by camptothecin or the flavonoid flavone via EMOD induced apoptosis, blocked by vitamin C (Wenzel et al, 2004). ==

Giant tumor cells of bone killed by EMOD induced apoptosis (Nicholson et al, 1998). =

Maxillary cancer of the nasal cavity killed or reversed by hydrogen peroxide (Sasaki et al, 1967). =

CT26 (colon) and Hepa 1-6 (liver) tumor cells killed by SOD mimic and EMOD induced apoptosis, blocked by NAC (Laurent et al, 2005). ==

Human neuroblastoma cells, SH-SY5Y, killed by EMOD induced apoptosis, blocked by thioredoxin (Andoh et al, 2003). ==

Human hepatoma PLC/PRF/5 cells killed by cinnamaldehyde via EMOD induced apoptosis, blocked by vitamin E (Wu et al, 2004). ==

B-cell Chronic lymphocytic leukemia (CLL) lymphocytes and acute myeloid leukemia (AML) blasts killed by adaphostin via EMOD induced apoptosis, blocked by NAC (Shanafelt et al, 2005). ==

HT-29 human colon cancer cells killed by alpha lipoic acid via EMOD induced apoptosis, blocked by benzoquinone (Wenzel, 2005). ==

Human H460 lung cancer cells killed by Bortezomib via EMOD induced apoptosis, blocked by Tiron (Ling et al, 2003). ==

Endometrial carcinoma cells killed by bortezomib via EMOD induced apoptosis, blocked by Tiron and vitamin C (Llobet et al, 2008). ==

Human colon carcinoma cell, including MIP101, DLD2, and HT29, suppressed by mitochondrial frataxin via EMOD induced apoptosis (Schulz et al, 2006). =

Human leukemia cell lines Jurkat and MOLT-4 killed by cannabidiol via EMOD Induced-apoptosis, blocked by ROS scavengers or the NAD(P)H oxidase inhibitors (McKallip et al, 2006). ==

Human breast cancer cells (MDA-MB-231 and MCF-7) killed by isothiocyanates via EMOD induced apoptosis, blocked by SOD and CAT (Xiao et al, Mol Cancer Ther. 2006). ==

MCF-7 and T47D breast cancer cell EMOD apoptosis from tamoxifen, blocked by vitamin E (Peralta et al, 2006). ==

HL-60 human leukemia cells killed by Dp3-Sam via EMOD induced apoptosis, blocked by NAC (Hou et al, 2005). ==

Human bladder cancer KU-1 cells killed by sic-dichlorodiamineplatinum via EMOD induced apoptosis, blocked by NAC (Miyajima et al, 1999). ==

U251 human glioblastoma cells killed by radiation via EMOD induced apoptosis, blocked by SOD and CAT (Lee et al, 2004). ==

HepG2 human hepatoma cells killed by EMODs (Paik et al, 2005). =

Human hepatoma cell lines PLC/PRF/5, H460, and HepG2 killed by etoposide via EMOD induced apoptosis, blocked by methalothionein (Shimoda et al, 2003). ==

Arsenic trioxide (As_2O_3) inhibited multiple myeloma cells, blocked by glutathione (Miller et al, 2002) (Dai et al, 1999). ==

Human leukemia cells PLB-985 killed by EMODs, blocked by CAT (Hiraoka et al, 1998). ==

Prostate cancer cells PC-3 killed by hyperthermia EMOD induced apoptosis, blocked by MnSOD overexpression (Venkataraman et al, 2008.) =

Stage IV metastatic melanoma, suppressed by Elesclomol (formerly STA-4783), produced oxidative EMOD apoptosis: blocked by NAC ==

Synta anticancer mechanism: STA-4783 causes EMOD apoptosis, blocked by NAC ==

Differential effect of ascorbic acid and NAC on As_2O_3-mediated oxidative stress in

Human leukemia (HL-60) cells, blocked by **NAC** (Yedjou et al, 2008). ==

<u>**Summary of Human Cell Lines Killed by EMOD Induced Apoptosis:**</u> **bladder, breast, colon, giant tumor cell of bone, glioblastoma, glioma, head and neck, histiocytic lymphoma, leukemia, lung, leukemia, liver, lymphocytic leukemia, lymphoma, maxillary cancer of nasal cavity, melanoma, metastatic melanoma, multiple myeloma, neuroblastoma, ovarian, pancreatic, prostate, renal cell. (23)**

Add to This:
- **human non-small cell lung cancer** (Ling et al, 2003) (Wu et al, 2006)
- **human colon adenocarcinoma** (Wenzel et al, 2005)
- **human colorectal carcinoma** (Chen et al, 2004) (Gali-Muhtasib et al, 2008)
- **human hepatocellular liver carcinoma** (Shimoda et al, 2003)
- **human acute myeloid leukemia** (Kay, 2006) (Chandra et al, 2003) (Shanafelt et al, 2005) (Mow et al, 2002)
- **human promyelocytic leukemia** (Hou et al, 2005)
- **human erythromyeloid leukemia** (Wagner et al, 2000)
- **human epithelial cancer cells** (breast and colon) (Aykin-Burns et al, 2009)
- **human endometrial cancer** (Llobet et al, 2008)
- **human invasive bladder cancer** (Miyajima et al, 1999 - human bladder cancer KU-1 cell line)
- **human osteosarcoma** (Ahmad et al, 2005)

This Makes a Total of **34 Human Cell Types** Killed by EMOD Induced Apoptosis.

Summary: 34 Human and 10 Murine Cancer Cell Lines Killed by EMOD-Induced Apoptosis

This makes a grand total of 34 human cancer cell lines and 10 murine cancer cell lines that have been shown to be killed by EMOD induced apoptosis, many of which are shielded or blocked by antioxidants.

MURINE

- Six murine tumor cell lines (**TLX9, P388, YAC, P815, J774** and **NK lymphoma cell lines**) killed by EMOD induced apoptosis, blocked by **glutathione and selenium** (Nathan et al, Vol 153, 1981). ==
- **Vitamin E** slows rate of free radical-mediated lipid peroxidation in **L1210** murine leukemia cells (Wagner et al, 1996). ==
- Murine pheochromocytoma (**PC12**) cells killed by EMOD induced apoptosis, blocked by **resveratrol** (Jang, Surh, 2001). ==
- Murine **Sarcoma 180** solid tumor cells killed by PEG-DAO via EMOD induced apoptosis (Fang et al, 2008). =
- Murine Melanoma **B16F10** and for Walker carcinosarcoma (**W256**) cells killed by EMODs (Jaganjac et al, 2008). =

- Murine B16F10 melanoma cells killed by TNP-470 EMOD induced apoptosis (Okroj et al, 2006). =
- Murine B16F10 melanoma cells killed by the fumagillin analog, TNP-470 EMOD induced apoptosis, blocked by NAC (Okroj et al, 2005). ==
- Murine thymoma cells WEHI7.2 killed by EMOD induced apoptosis, blocked by CAT over-expression (Tome et al, 2001). ==

MURINE CELL LINES SUMMARIZED:
Six Murine Tumor Cell Lines (TLX9, P388, YAC, P815, J774 and NK lymphoma cell lines)
L1210 murine leukemia cells
Murine pheochromocytoma (PC12) cells
Murine Sarcoma 180 solid tumor cells
Murine Melanoma B16F10 and for Walker carcinosarcoma (W256) cells
Murine thymoma cells WEHI7.2
Add to This:
  - **murine mammary cancer** (Bracke et al, 1999)
  - **murine brain cancer** (Zeisel (2), 2004)
  - **murine fibrosarcoma** (Teicher et al, 1994)
  - **murine retinoblastoma** (Salganik et al, 2000)

This makes a total of **10 murine cell types** killed by EMOD induced apoptosis, some of which can be blocked by antioxidants. Also presented is **52 instances EMOD induced apoptosis being blocked by antioxidants, resulting protection of the cancer cells.**

According to the American Cancer Society, in 2016-17 there are over 15.5 million cancer survivors in the United States of America. If these individuals are taking antioxidants, this may represent a huge potential public health problem, whereby these survivors are endangering their current and future wellbeing, by unintentionally sheltering cancer cells secondary to excessive antioxidant use.

- Human hepatoma PLC/PRF/5 (CD95-negative) cells Phenylethyl isothiocyanate (PEITC) apoptosis, blocked by NAC and vitamin E (Wu et al, 2005). ==

Phenylethyl isothiocyanate (PEITC) is a well-recognized potential chemopreventive compound against human cancers. In this study, the molecular mechanism of PEITC-induced apoptosis was examined with two antioxidants (N-acetyl-cysteine and vitamin E) and a caspase-3 inhibitor (z-DEVD-fmk). Results demonstrated that PEITC significantly induced **human hepatoma PLC/PRF/5** (CD95-negative) cells undergoing apoptosis. Treatment with 0 approximately 10 microM PEITC-triggered cell apoptosis as revealed by the externalization of annexin V-targeted phosphatidylserine and the subsequent appearance of sub-G1 population. Results also displayed that **PEITC-induced apoptosis** involves the up-regulation of p53 and Bax protein, down-regulation of the XIAP, Bcl-2, Bcl-(XL) and Mcl-1 proteins, cleavage of Bid, and the release of cytochrome c and Smac/Diablo, which were accompanied by the activation of caspases -9, -3 and -8. **PEITC-induced the generation of reactive oxygen species** and the decrease of mitochondrial membrane potential (Deltapsim) in a time-dependent pattern. **N-acetyl-cysteine and vitamin E at 100 microM, and z-DEVD-fmk at 50 microM markedly blocked PEITC-induced apoptosis, which was demonstrated by a decline in the reactive oxygen species generation** and the release of the cytochrome c and Smac/Diablo from mitochondria to the cytosol. N-acetyl-cysteine, vitamin E and z-DEVD-fmk also prevented the PEITC in inducing the loss of Deltapsim. They also affected the activity of XIAP and Bax proteins. Taken together, these studies suggest that PEITC is an apoptotic inducer that acts on the mitochondria and the feedback amplification loop of caspase-8/Bid pathways in PLC/PRF/5 cells. (Wu et al, 2005).

RMH Note: Clearly, human cancer cells will be allowed to proliferate by the use of NAC and vitamin E.

RMH Note: Conclusions drawn from these studies suggest that there is a higher prevalence of antioxidant dietary supplement use among cancer survivors than among the general population, and that supplements being used are an increasingly complex mixture of ingredients.

RMH Note: I believe that this is a very dangerous practice, since it has repeatedly been shown that antioxidants block EMODs ability to kill cancer cells.

Cancer Cells Know How to Protect Themselves

Lung cancer patients have higher GPx and lower SOD levels as determined by alterations of antioxidant activities in erythrocytes from patients with non-small cell lung carcinoma (NSCLC). In total, 189 cases of mostly advanced-stage IIIB or stage IV NSCLC and 202 healthy controls were studied. In subjects with lung cancer, there was similar catalase activity, **lower SOD activity** and **higher GPx** compared with controls. However, **more advanced disease (stage IV compared with stage IIIB) was associated with lower SOD activity**. Non-small cell lung carcinoma in Chinese subjects is associated with alterations in systemic antioxidant activities, which may play an important role in carcinogenesis and progression of cancer cell growth. (Ho et al, 2007).

RMH Note: I predicted that lung cancer patients would have higher GPx levels and lower SOD levels. The GPx would lead to an EMOD insufficiency and the low SOD would produce lowered amounts of tumoricidal peroxide. Both situations are conducive to allowing tumor growth. I feel that this goes hand in hand with the cancer cells to protect itself from EMOD induced apoptosis by having high levels of the antioxidants, vitamin C, lactate and glutathione.

We also need to discuss the antioxidant, lactic acid.

RMH Note: Lactic acid is an antioxidant and it is in this capacity that it is advantageous for cancer cells to produce it, such that EMODs can be kept at sub-apoptotic levels. Salts of lactic acid are used in foods as a humectant and an antioxidant which can increase the effect of other antioxidants.

My Summary of Cancer Cell Types Killed by EMODs (ROS) (From Gupta et al, 2011)

H_2O_2 causes apoptosis in:
hepatoma cells, leukemia cells, osteosarcoma, breast, bladder, and lung cancer cells. Also procarbazine produced peroxide that killed Hodgkin's lymphoma, non-Hodgkin's lymphoma and primary brain tumors.

ROS (EMODs) induced apoptosis occurs in:
chronic myeloid leukemia, lymphoma cells, and prostate cancer cells.

ROS (EMODs) induce autophagic death occurs in:
breast cancer, non-small cell lung cancer (NSCLC), glioma, neuroblastoma, glioblastoma, and cervical cancer.

ROS (EMODs) generated by gemcitabine and by thymoquinone have also been shown to inhibit the growth of pancreatic and prostate cancer cells.

As_2O_3 has the ability to induce superoxide production in cancer cells and treat acute promyelocytic leukemia.

Some cancer types for which ROS (EMODs) have been shown to play a role in radiation-induced cancer cell death are lung adenocarcinoma, non-small-cell-lung cancer, prostate cancer, and breast cancer.

Other clinical studies for which ROS have been shown to play a role in radiation-induced therapy include patients with head and neck squamous cell carcinoma, cervical cancer, prostate cancer, NSCLC, rectal cancer, and breast cancer.

**Certain self-evident truths
need not require additional evidence or proof,
such as the low toxicity of EMODs,
as evidenced by their omnipresent ubiquity.**
R.M. Howes, M.D., Ph.D.
7-15-09

**If free radicals were toxic,
we would all be dead.**
R.M. Howes, M.D., Ph.D.
3-19-18

Antioxidants have failed to block, reverse or prevent diseases, such as cancer, strokes and cardiovascular disease, in randomized clinical trials with over 16 million participants and clearly show that the free radical theory lacks predictability and had been invalidated because it fails to meet the requirements of the scientific method. I have researched this extensively (Howes RM. Death in Small Doses: Book One, 2010) (Howes RM. Death in Small Doses: Book Two, 2010) (Howes, Linked, 2012).

In actuality, free radicals perform a crucial role in normal, healthy physiological processes like our immune system and promote (prooxidative reactions) beneficial oxidation.

It is important to realize that **many vitamins and supplements classified as antioxidants (or so-called antioxidants) are actually *redox agents*, meaning they act as antioxidants in some instances and prooxidants in others. This markedly increases the difficulty of interpreting redox data and is seldom addressed in the literature.**

Basically, **oxygen free radicals are of low toxicity and are crucial metabolic agents for pathogen and neoplasia protection. The antioxidant study failures were due to the fact that they were based on a debunked and out-dated theory** (Howes, Philica. Feb 26, 2007).

Please keep in mind that the apoptotic process is applicable to all cells and not just to cancer cells. It serves as a mechanism to rid the body of damaged, aged or mutated cells.

To summarize, increased mortality, cancer, heart disease and strokes are linked to antioxidants.

- Dietary supplements are a $28 billion industry and they are taken by 53% of the US population. Medical personnel and cancer survivors take more than the average person.

- No supplements are checked for effectiveness or safety by governmental agencies. According the DSHEA of 1994, dietary supplements are "foods" and not medicines.

- A JAMA 2007 article on nearly 250,000 subjects showed that the antioxidant vitamins A, C and E increase the incidence of cancer, heart disease and strokes and overall mortality by 5%.

- Three of the largest studies (RCTs) were shut down 21 months early because the supervisory physicians felt it was unethical to continue and place participants in harm's way. (ATBC, CARET, SELECT). Almost 90,000 subjects were involved.

- Only in cases of proven vitamin deficiency or malabsorption syndrome should these agents be taken. Those in areas of poor nutrition may also consider them.

- Chemically concocted supplements do not act with the same beneficial effects in the human body. Combining the antioxidant vitamins only increases their adverse effects.

- The 5 most common natural antioxidants cause serious harm when present in excessive amounts: cholesterol (heart disease), uric acid (gout and heart disease), estrogen (uterine and breast cancer), testosterone (prostate cancer), and bilirubin (jaundice and brain damage).

- Supplements are prohibited from using the four words: diagnose, prevent, cure or treat. Otherwise, any other words can spin the products, such as strengthen, support, bolster, improve, fortify, build up, etc.

- 32 major medical/scientific organizations do not recommend taking antioxidant supplements. (FDA, AHA, ACS, NCI, ACCP, USPSTF, etc.).

- I have collected over 500 study reports showing the ineffectiveness of the antioxidants and of these, 170 studies show harmful consequences. This must stop!!!

- The best way to get vitamins is to eat foods that naturally contain them.

MORTALITY AND EXCESS ANTIOXIDANTS

Google headlines have addressed the increased mortality that has been linked to antioxidants as follows: (accessed 2-26-18)

JAMA Meta-Analysis: Antioxidants Increase Mortality

www.naturalproductsinsider.com/.../02/...**antioxidants-increase**-m.aspx

COPENHAGEN, Denmark—A meta-analysis out of Copenhagen University Hospital found the use of **antioxidant** dietary supplements—specifically beta-carotene,

Antioxidant Vitamins_ Review Finds Increase in Mortality ...

https://www.scribd.com/document/216724567/**Antioxidant**-Vitamins...

Antioxidant Vitamins_ Review Finds Increase in **Mortality** - Free download as PDF File (.pdf), Text File (.txt) or read online for free.

Confirmed: vitamin pills can cause death | ScienceNordic

sciencenordic.com/**confirmed-vitamin-pills-can-cause-death**

Confirmed: vitamin pills can cause death. ... The analyses show that the **antioxidants increase mortality** significantly with a comparative risk **increase** of four ...

Vitamins A, C and E Increase Mortality ... - naturalnews.com

https://**www.naturalnews.com**/023034_vitamins_**antioxidant**_junk...

Scientists finally issue warning against canola oil: Study reveals it is detrimental to brain health, contributes to dementia, causes weight gain - **NaturalNews.com**

Research: Vitamins may increase risk of death - CNN.com

www.cnn.com/2008/HEALTH/diet.fitness/04/16/vitamins.health/index.html

Apr 16, 2008 · The experts said the studies involved different doses of each antioxidant. A total of 232,550 people were involved. Forty-seven trials included 180,938 people and had a low risk of bias. In these trials with a low risk of bias, the "antioxidant supplements significantly increased mortality", the report's authors wrote.

Antioxidant Supplements Seem to Increase Mortality ...

https://**www.biospace.com**/article/around-the-web/**antioxidant**...

Antioxidant Supplements Seem to **Increase Mortality**, Cochrane Review Shows - read this article along with other careers information, tips and advice on BioSpace

Antioxidant supplements tied to death risk | Reuters

https://**www.reuters.com**/article/us-vitamins-hazards/**antioxidant**...

Feb 27, 2007 · Antioxidants are believed to fight free radicals, atoms or groups of atoms formed in such a way that they can cause cell damage. "Beta carotene, vitamin A, and vitamin E given singly or combined with other antioxidant supplements significantly increase mortality," the study found.

Antioxidant Vitamins May Increase Mortality - Medscape

https://**www.medscape.org**/viewarticle/552910

Antioxidant supplements for prevention of mortality in ...

www.**cochrane**.org/CD007176

In the 56 trials with a low risk of bias, the **antioxidant** supplements significantly increased **mortality** (18,833 dead/146,320 (12.9%) versus 10,320 dead/97,736 (10.6%); RR 1.04, 95% CI 1.01 to 1.07). This effect was confirmed by trial sequential analysis.

Use Of Some Antioxidant Supplements May Increase Mortality ...

Study Showing Antioxidant Vitamins Increase Mortality ...

www.highlighthealth.com/research/study-showing-**antioxidant**...
A study published in The Journal of the American Medical Association (JAMA) made headlines recently. The review, "**Mortality** in randomized trials of antioxi

Use of Some Antioxidant Supplements May Increase Mortality ...

https://www.mybesthealthportal.net/nutrition/vitamins-and...
Contradicting claims of disease prevention, an analysis of previous studies indicates that the **antioxidant** supplements beta carotene, vitamin A, and vitamin E may ...

Study links antioxidant supplements to increased mortality

https://www.nutraingredients.com/Article/2007/02/28/Study-links...
A meta-analysis of 68 randomised trials with **antioxidant** supplements has reported that vitamins A and E, and beta-carotene may **increase mortality** risk by up to 16 per ...

Mortality in randomized trials of antioxidant supplements ...

https://**www.ncbi.nlm.nih.gov**/pubmed/17327526
Feb 28, 2007 · DATA SYNTHESIS: When all low- and high-bias risk trials of antioxidant supplements were pooled together there was no significant effect on mortality (RR, 1.02; 95% CI, 0.98-1.06). Multivariate meta-regression analyses showed that low-bias risk trials (RR, 1.16; 95% CI, 1.04[corrected]-1.29) and selenium (RR, 0.998; 95% CI, 0.997-0.9995) were significantly associated with mortality.

Both the 1994 and 1996 ATBC studies on 29,133 participants showed an increased incidence and mortality of lung cancer with use of alpha-tocopherol and B-carotene.

And the 1996 and 2004 CARET studies on 18,314 participants found an increased mortality associated with vitamin A and B-carotene antioxidant supplementation.

A meta-analysis in 2004 on 131,727 participants found that B-carotene, vitamin A, vitamin E and selenium antioxidant treatment did not prevent gastrointestinal cancer but significantly increased mortality.

Also, the Iowa Woman' Health Study in 2011 on 38,772 participants found that vitamin and mineral supplements maybe associated with increased total mortality risk.

Food for Thought

We have all been told that consuming antioxidants will make us look younger, feel younger, boost our immunity, fight diseases (i.e., cancer, heart disease, strokes, diabetes, arthritis), and increase our longevity. But, we have been misled...intentionally misled!

Since supplements tend to be most perceived for their apparent medicinal qualities, that is where the problems start. They are classified as "food" and not drugs.

The FDA will not remove a supplement from the market until it has been shown to be harmful to people and there is a lack of a program to report adverse effects of these agents and according to Michael McCann, the FDA fails to learn of over 99 percent of adverse consumer reactions. Thus, people serve as guinea pigs until it becomes obvious that these supplements are harmful. I believe that time is now.

Actually, the fact is that the FDA has all the legal authority it needs to remove supplements that contain illegal drugs or have been shown to be harmful from the market. The FDA has failed to do its job, and there are companies selling dietary supplements that contain prescription drugs and I have found hundreds of studies showing the ineffectiveness and harmful potential of the antioxidant vitamins.

The "supplement loophole" in the FDA's regulatory authority needs to be corrected and the FDA should serve to protect ever gullible citizens from the potential risk and old-fashioned fraud by the hucksters of unproven and dangerous products. Deliberately fraudulent health claims are everywhere. We now have the most sophisticated snake oil salesmen of all times.

The following was taken from: **Office of Dietary Supplements, National Institutes of Health**. http://ods.od.nih.gov/pubs/DS_WhatYouNeedToKnow.pdf. Accessed 12-14-10.

"Many supplements contain active ingredients that can have strong effects in the body. Supplements are most likely to cause side effects or harm when people take them instead of prescribed medicines or when people take many supplements in combination. Some can increase the risk of bleeding or they can affect a person's reaction to anesthesia or interact with certain prescription drugs in harmful ways. Antioxidant supplements, like vitamins C and E, might reduce the effectiveness of some types of cancer chemotherapy. Keep in mind that some ingredients found in dietary supplements are added to a growing number of foods, including breakfast cereals and beverages. As a result, you may be getting more of these ingredients than you think, and more might not be better. Taking more than you need is always more expensive and can also raise your risk of experiencing side effects."

Ironically, this NIH site also makes this statement, "The federal government can take legal action against companies and Web sites that sell supplements when the companies make false or deceptive statements about their products, if they promote them as treatments or cures for diseases, or if their products are unsafe."

It is readily apparent that false and deceptive statements are everywhere, especially when they claim medicinal qualities.

The 21st century has become the golden age of dietary supplement quackery.

What's a conscientious person to do? Pop more vitamin pills and go on your way?

Not so fast, says Howard Sesso, ScD, MPH, an associate professor of medicine at Harvard Medical School and project director of the Physicians' Health Study II, which followed nearly 15,000 male doctors over the course of 10 years to evaluate the potential health benefits of four of the most popular vitamins: C, E, beta carotene, and multivitamins.

Sesso told WebMD, "Studies over the last few years have suggested that individual supplements don't have benefits."

In one large study, Sesso and his fellow researchers found that neither vitamin E or C lowered the risk of cardiovascular disease. That study was published in *The Journal of the American Medical Association (JAMA)* in November 2008. In a study published in JAMA in January 2009, the same team reported that taking supplements of vitamins E and C did not lower a man's risk of developing prostate or total cancers.

"You have to look at the totality of evidence in order to come up with recommendations," says Andrew Shao, PhD, senior vice president of scientific and regulatory affairs at the Council for Responsible Nutrition, a trade organization that represents the nearly $27 billion-a-year supplement industry (figures vary and go all the way $23 to $35 billion). They also say that, "There is no magic bullet."

Norman Krinsky, Ph.D., chair of the Institute Of Medicine's Panel on Dietary Antioxidants and Related Compounds, has concluded, "There has been much confusion about the value of taking antioxidant supplements. **Many people are likely to continue to believe that these compounds have many health benefits, including the prevention of chronic disease, but "we were not able to find definitive proof of that hypothesis."** Dr. Krinsky is a professor of biochemistry at Tufts University Boston.

Serious medical damage, and potentially death, may be averted only if the patients taking excessive antioxidant supplements stop the drugs at the first sign of untoward effects. Likely, the first sign will be lethargy or unexplained tiredness, as was described by Dr. Denham Harman in a discussion with Jack Challem. Even early stoppage may not guarantee avoidance of long term unintended consequences.

The following notes were excerpts of Jack Challem's recount of a conversation with Dr. Denham Harman, father of the nullified free radical theory:

With all the attention given to antioxidants, a lot of questions remain unanswered. How much is enough? How much is too much? Are free radicals really all that bad? Challem began asking these questions because the vitamins Challem took should have left him feeling energized. But over the past 10 years, he had felt far more tired than he should - tired in the morning, tired in the afternoon, tired in the evening.

*It wasn't easy finding answers. Challem finally tracked down the guy who would know: Denham Harman, MD, PhD. He's the fellow who invented the free radical theory of aging back in November of 1954. To Challem's surprise, he discovered that he was doing too good a job crushing those dangerous free radicals we all hear about. Challem **was taking too many antioxidants. With all the talk about the benefits of antioxidants, we forget that free radicals are there for a reason. Free radicals are essential for health.***

The relationship between free radicals and antioxidants is one of balance. Challem asked Harman which supplements he took. Unlike a lot of doctors who don't want to go on record with this information, he was very up front: 400 IU vitamin E, 2000mg vitamin C, 100 micrograms selenium and 30mg CoQ10 each day and 25,000 IU of beta-carotene every other day. **"I'd take more,"** *he said,* **"but I can't afford to get tired."** *Challem's ears perked up.* **He explained that too many antioxidants cause fatigue and muscle weakness.**

Several years ago, in an experiment, Harman found that large amounts of BHT, a man-made antioxidant, interfered with the ability of mice to produce energy. **"Too many antioxidants can leave you feeling very weak,"** *Harman said.* **"BHT decreases ATP and mitochondria function."** *ATP, or adenosine-tri-phosphate, is required for energy production in the mitochondria, which is the part of the cell biologists call the energy factory. Challem asked Harman whether too many natural antioxidants could cause fatigue.*

Harman was positive, "Yes!" **There's a point where the stuff turns on you.** *(Jack Challem. Some Good Things to Say About Free Radicals. Reproduced from The Nutrition Reporter® newsletter).*

RMH Note: Dr. Harman postulated the now debunked free radical theory and even he recognized the dangers of ingesting excessive amounts of antioxidant supplements.

The beauty of the Free Radi-Crap Theory of oxidative stress and aging (as I like to call it) is that it is testable, and it has miserably failed the test hundreds of times. Yet, sycophantic zealots cling tenaciously to its flawed precepts.

If the free radical theory was correct, antioxidants should and would prevent/reverse/cure all of the 100-200 diseases attributed to oxidative stress and stop/reverse aging. But, antioxidants have repeatedly failed to do so. They have been an overall disappointing failure.

Maureen Storey, Ph.D., a nutritionist with the Georgetown Center for Food and Nutrition Policy, stated, **"You can get too much of a good thing."** People who pop large amounts of vitamins or minerals--thinking that if a little is good, a lot is better--need to know that there are dangerous levels of these compounds.

In November 2004, the American Heart Association warned that while the small amounts of vitamin E found in multivitamins and foods were not harmful, taking 400 International Units a day or more could increase the risk of death.

But, please keep in mind that these vitamins can serve in many other important cellular functions, other than in oxidation-reduction reactions (redox reactions). Thus, one should avoid vitamin deficiencies, even of the antioxidant vitamins.

In its announcement proposing Medication Guides (patient package inserts) to provide prescription drug customers with comprehensive and reliable drug information, the FDA stated, "FDA believes that improved dissemination of information about prescription drug products is necessary to fulfill patients' need and right to be informed." Medication Guides are intended to be used in products "that pose a serious and significant public health concern" requiring immediate distribution of drug information to the public." (FDA, 1995).

Mistrust in our governmental agencies (FDA, EPA) and distrust of the entire pharmaceutical and dietary supplement industry is also a reality and a growing problem. Legally prescribed drugs are killing more people (over 106,000 per year) than illegal drugs and the deep pockets of drug companies allows them to dominate television and printed advertising.

Just consider these recent examples of harmful drug controversy: Accutane, Baycol, Vioxx, Celebrex, hormone replacement therapy, Darvon, Darvocet, Avandia, etc. The list goes on and on. Remember, these drugs were "tested" (supplements are not tested) and they can still have deadly consequences.

There is even growing mistrust of the medical sector, because they are seen as promoting drugs unnecessarily and making profits off of the drugs they prescribe, many of which are deadly. As Steve Covey said, "Common sense is not so common anymore."

Today, people are encouraged to run to the doctor and gulp down handfuls of all varieties of medications, many with serious side effects. However, whatever drugs doctors claim are good for you today are likely to change by tomorrow, such as Vioxx, Celebrex, Avandia, hormone replacement therapy, etc. It seems that what was good for you today, will be bad for you next week.

More and more it appears that drug companies, most of which are foreign owned, are only interested in the profit margins and returns for their stockholders. Drugs and vaccines are being recommended and sometimes required (mandated) for a younger and younger population, such as the HPV vaccine and cholesterol lowering drugs for those as young as 8 years old. Now, they are targeting babies and toddlers with unending vaccinations. But, you have got to admit that it opens up huge new drug markets for more of their profits.

Americans are the biggest group of "pill poppers" on the planet. A government report indicates that, for the first time, abuse of painkillers and other medication is sending as many people to the emergency room as the use of illegal drugs.

Rest assured, they can kill you.

I have addressed this very subject because, just like drugs, antioxidant supplements can also result in significant public health concern. The public must be informed of this and the manufacturers, purveyors, vendors or sellers of these products will not inform the public due to fear of hurting sales and incurring legal liabilities.

Wide ranging antioxidant agents which block oxidation (food industry (preservatives, rancidity, spoilage, browning); the rubber and plastics industry; pharmaceutical industry; antioxidants in beverages and herbal products; antioxidants in all processed foods; etc.) outside of the body have the capability to block oxidation inside of the body, if ingested and absorbed. There may be as many as 25,000 phytochemicals in the human diet, many having direct antioxidant effects. That is where the potential harm and dangers start with antioxidant use and overload in humans.

We must maintain sufficient electron-donating antioxidants in order to form adequate protective levels of electron-accepting prooxidants. Problems develop when antioxidant over dosing results in an electronically modified oxygen derivative (EMOD) insufficiency. I prefer to more accurately refer to oxygen free radicals as EMODs.

Specifically, the accumulating evidence presented in my books clearly identifies the antioxidant supplements as agents that pose a significant public health concern. To be sure, there are some non-randomized, non-blinded studies showing that the antioxidants can be beneficial, but we cannot ignore the over 500 studies showing adverse effects. With such large populations taking antioxidants on a daily basis, this truly is a global public health issue.

We must base our decisions regarding ingestion of antioxidant supplements on the best overall scientifically based evidence, even though there is no over-riding, monolithic, absolute conclusion, which can be supported by all of the literature.

We should follow the recommendations of the Mayo Clinic. "The U.S. Food and Drug Administration does not strictly regulate herbs and supplements. There is no guarantee of strength, purity or safety of products, and effects may vary. You should always read product labels. If you have a medical condition, or are taking other drugs, herbs, or supplements, you should speak with a qualified healthcare provider before starting a new therapy. Consult a healthcare provider immediately if you experience side effects." (http://www.mayoclinic.com/health/echinacea/NS_patient-echinacea/DSECTION=safety (accessed 12-25-10).

Misinformation and false impressions have led to bad decisions by millions of consumers. Although many researchers and clinicians have suggested that vitamin supplements, especially high-dose antioxidants, should not be used by patients during cancer treatment, vitamin supplement use is widespread amongst most cancer patients.

Studies indicate that antioxidants can block tumoricidal therapies such as chemotherapy and radiotherapy. If antioxidants protect tumor cells to any extent, they could theoretically lower the efficacy of cancer treatments and promote cancer cell growth, metastasis, recurrence and death.

Even Multivitamins are Being Questioned

Scientists at the Fred Hutchinson Cancer Research Center followed 160,000 postmenopausal women for about 10 years. Marian Neuhouser, PhD., lead author, concluded: "Multivitamins failed to prevent cancer, heart disease, and all causes of death for all women. Whether the women were healthy eaters or ate very few fruits and vegetables, the results were the same."

Miriam Nelson, PhD, director of the John Hancock Research Center on Physical Activity, Nutrition, and Obesity at Tufts University, said, "The multivitamin as insurance policy is an old wives' tale, and we need to debunk it." This represents a complete reversal of the thinking surrounding our health care policies, because as recently as 2002, no less an authority than the

Journal of the American Medical Association recommended that "all adults take one multivitamin daily."

Shockingly, as of November 2010, Prevention magazine said that studies show that it may be "Time to kick the multivitamin habit." Studies on boosting immunity have been "equally discouraging" and a British review of eight studies found no evidence that multivitamins reduced infections in older adults.

Other studies even indicate that multivitamins may be harmful. A 2010 study of Swedish women found that those who took multivitamins were 19% more likely to be diagnosed with breast cancer over a 10-year period than those who didn't. Other studies have shown them to cause increased breast density in women, which is associated with breast cancer.

A 2007 paper, in the Journal of the National Cancer Institute, found that men who took multivitamins along with other supplements were at increased risk of prostate cancer, especially the fatal type of prostate cancer.

And other research has linked excessive folic acid intake to higher colon cancer risk in people who are predisposed. David Katz, MD, MPH, director of the Prevention Research Center at Yale University School of Medicine, said, "In terms of a risk-benefit ratio, **why would you accept even a tiny risk if you're not getting any benefit?"**

Excellent question!

What is **the wisdom of taking supplements without a clear understanding of their potentially harmful side effects?**

Unfortunately, consumers presume these supplements are safe and use them without their doctors' supervision. Of even more concern, many doctors are barely knowledgeable of the harmful potential of the antioxidant vitamins. Many **practitioners argue that their remedies are effective, despite overwhelming scientific evidence to the contrary**.

According to the National Institutes of Health, more than half of adults in the U.S. consume some kind of antioxidant product, spending $37 billion each year.

Although some people spend countless dollars on antioxidant supplements to improve their health, many studies have found that these would-be panaceas could actually exacerbate the diseases they claim to prevent.

Some of the following was modified or excerpted from:

THE DOWNSIDE OF ANTIOXIDANTS

The following are a few facts I have uncovered in the scientific literature that will shock you.

ANTIOXIDANT FACTS:
- Antioxidants have increased mortality (death) by as high as 17% and 19%.

- Antioxidants have increased lung cancer by 18% and 28%, esophageal cancer deaths by 14% and 22.1%, breast cancer by 19%, hemorrhagic stroke deaths by 50%, ischemic heart disease by 11%, and cardiovascular disease by 18%. Prostate cancer risks increased 17%.

- Antioxidants have increased prostate cancer deaths, elevated the risk of squamous cell carcinoma, doubled the risk of adenoma recurrence, increased the rate of second primary cancers, and increased recurrence of head and neck tumors.

- Antioxidants have increased ischemic heart disease, deaths from fatal coronary heart disease, increased risk of nonfatal and fatal myocardial infarction, decreased platelet function, increased intima-to-media thickness, negated statin effects, increased risk of hospitalization for heart failure and hypertension, altered liver function tests, increased blood pressure, and increased blood loss after cardiac surgery.

- Antioxidants have increased hemorrhagic stroke deaths as much as 22% to 50%, increased risk of subarachnoid hemorrhage 50% and increased risk of intracerebral hemorrhage 62% and increased risk of fatal subarachnoid hemorrhage 181%.

- Antioxidants have increased wheezing, productive coughs, and risk of asthma.
- Antioxidants have increased risk of tuberculosis by 72% and pneumonia by 14%, indicating an altered immune system.

- Antioxidants have adversely affected muscle performance and hampered endurance capacity.

Unbelievably, my research has found twenty-seven (27) types of human cancer cell types and nine (9) murine (rodent) cancer cell types that can be killed by EMODs (electronically modified oxygen derivatives) and in which the killing can be blocked by antioxidants, thereby providing antioxidant protection and shielding of the cancer cells.

To anyone who feels guilty for not gorging on antioxidants—actually, make that "antioxidants!" which seems to be how grocery manufacturers think of them—redemption is near. **(Begley, 2001)**

For years the media, food labels, dietitians, and even scientists who should know better have bombarded us with advice to load up on antioxidants: compounds found (mostly) in fruits and vegetables that allegedly mop up free radicals, which are supposedly highly reactive clusters of atoms that have been fingered as the evildoers responsible for aging and for illnesses from cancer to heart disease.

But, not so fast. First, studies piled up showing that taking antioxidants—even such common and seemingly innocuous ones as beta carotene and vitamins C and E—as supplements was not beneficial to health and might even be dangerous, though the reason for the danger wasn't clear.

One always pays attention when a study concludes with a phrase like "seems to increase overall mortality."

Now the research is challenging an even more fundamental tenet of the antioxidant craze. Many of the free radicals that are neutralized by antioxidants perform valuable functions in the body.

The most important: fighting toxins (white blood cells churn out free radicals by the battalion to fight bacterial infection) and fighting cancer. Maybe it's not such a fabulous idea to flood the body with something that neutralizes these warriors of the immune system.

Or as British chemist and science writer David Bradley noted in his blog, Reactive Reports, "It's always struck me as odd that you would want to ingest extra antioxidants anyway, given that oxidizing agents are at the front-line of immune defense against pathogens and cancer cells ... Suffice to say that taking antioxidant supplements ... may not necessarily be good for your health if you already have health problems," especially cancer or an infection.

The first hints that the bandwagon was crashing came from the hundreds of studies that have tried to assess the health effects of antioxidant supplements.

The results have not been pretty. In 2008 the Cochrane Collaboration, an international consortium of scientists who assess medical research, scrutinized 67 studies with nearly 400,000 participants. The goal: to determine whether antioxidant supplements reduce mortality in either healthy people or in people with cardiovascular, neurological, rheumatoid, renal, endocrine, or other diseases. Conclusion: "We found no evidence to support antioxidant supplements for primary or secondary prevention, [and] Vitamin A, beta-carotene, and vitamin E may increase mortality."

In analyses of antioxidant supplements and Lou Gehrig's disease, Alzheimer's or mild cognitive impairment, and lung cancer, the Cochrane scientists' verdict was the same: no, no, no, and no. And each analysis had an alarming refrain about increasing overall mortality.

It's not clear to other scientists why antioxidants in supplement form might be so dangerous.

One idea holds that at high doses they become pro-oxidants, stimulating the harmful DNA- and cell-damaging reactions they're supposed to prevent. But a more likely explanation is that we are seeing the human version of what scientists are finding in studies of lab animals: antioxidants interfere with immune-system cells that fight infection and cancer.

For those who got to this party late, free radicals are generated by normal metabolism, though dietary fat and iron-rich foods such as red meat generate more of them.

In the laboratory, free radicals can damage proteins, fats, carbohydrates, and DNA—the biological stuff we're made of. They can also harm the cells that line blood vessels, enabling tumor cells to enter the bloodstream and metastasize.

All this destruction, erroneously explains Jeffrey Blumberg, director of the Antioxidants Research Lab at Tufts University, "is believed to initiate, promote, or stimulate the progression of many chronic diseases, including cardiovascular disease and cancer," as well as normal aging.

"The hypothesis that emerged is that if you prevent the damage, you prevent the disease." This is same old outdated cry of "antioxidants to the rescue."

That was music to the ears of food manufacturers, especially as research suggested that some antioxidants are better than others at quenching different free radicals. Beta carotene, for instance, slays superoxide dismutase, but vitamin E is relatively impotent against it.

On the other hand, vitamin E makes LDL ("bad" cholesterol) more resistant to oxidation, which should be good for health: oxidized LDL is more likely to form plaque deposits that clog arterial walls. The different functions of different antioxidants opened the door to marketers pushing supplements for each one, as well as otherwise-unhealthy foods fortified with a cornucopia of antioxidants (Chocolatey Peanut Butter FiberPlus Antioxidants, anyone?).

In addition to the alarming findings from studies of people who take antioxidant supplements, new research is casting serious doubt on the benefits even more directly:

• A paper to appear in an issue of the *Proceedings of the National Academy of Sciences* finds that antioxidants might impair fertility. When scientists led by developmental biologist Nava Dekel of Israel's Weizmann Institute of Science applied antioxidants to the ovaries of female mice, ovulation levels plummeted: follicles released very few eggs. That suggested that ovulation might require the free radicals that antioxidants neutralize. Further experiments confirmed it: a type of free radical called reactive oxygen species is produced in response to luteinizing hormone, the physiological trigger for ovulation. That suggests that luteinizing hormone triggers ovulation through an intermediary—namely, reactive oxygen species/EMODs. **If reactive oxygen species are being mopped up by antioxidants, there's no ovulation.**

• A 2010 study in lab rats found that two popular antioxidants, quercetin (found in black and green tea, red onion, and other plant foods) and ferulic acid (in apples, artichokes, wheat, and other plants), aggravated and possibly triggered kidney cancer. As the scientists put it in the *Journal of Agricultural and Food Chemistry*, **"It is time to reevaluate the tumorigenic detrimental effect of" antioxidants."**

• Finally, a recent study in lab mice finds that a natural protein that boosts antioxidant levels in the blood may actually promote atherosclerosis, or clogging of the arteries. The study, in the issue of *Arteriosclerosis, Thrombosis, and Vascular Biology,* offers clues about why taking antioxidants has not been shown to improve heart health. The protein Nrf2 indeed boosts antioxidants, but in the study it also raised blood-cholesterol levels, as well as cholesterol content in the liver—both of which are excellent ways to get atherosclerosis.

In 2009, 108 new food products with antioxidants touted on the label reached store shelves in the United States, according to the market-research firm Mintel. That compares with 16 in 2005 and 82 in 2007. "Buyer beware" doesn't begin to cover it.

I have personally been one of the most dominant authors against excessive antioxidant ingestion for decades. This is evidenced by my books.

My books are my attempts to communicate the potential and known dangers of antioxidants to the medical community and to the overall populous. Also, they are my way of introducing people to my attempts at revolutionizing cancer and heart disease prevention and cure, all based on oxygen derivatives (EMODs).

Some of my books are available on www.amazon.com as follows:

[Death in Small Doses? Trafford Publishing, © 2010. 348 pages]

[Antioxidant Overkill. © 2011. 428 pages]

[Dangers of Excessive Antioxidants in Cancer Patients. © 2011. 514 pages]

[Heart Disease and Antioxidant Failures. © 2011. 634 pages]

[Antioxidant Failures and Dangers. © 2011. 434 pages]

[Anti-Aging Anti-Oxidant Scams. © 2011. 332 pages]

[Sports, Athletes, Exercise Facts and Antioxidant Myths. © 2011. 422 pages]

[Alzheimer's Disease: Forget Antioxidants and Supplements. © 2012. 362 pages]

[Sex, Performance, Reproduction, Naked Radicals and Antioxidants. © 2012. 310 pages]

[Antioxidants Linked to Deadly Unintended Consequences. © 2013. 341 pages]

[U.T.O.P.I.A.: Unified Theory of Oxygen Participation in Aerobiosis. © 2014, revised. 518 pages]

[Hydrogen Peroxide: A Health, Homeostatic and Protective Essentiality. © 2014. 492 pages]

[Reactive Oxygen Species vs. Antioxidants: The Oxypocalypse or The War That Never Was. © 2014. 670 pages]

[Diabetes and Oxygen Free Radical Sophistry. © 2014, revised. 434 pages]

[FISH OIL [Omega3 fatty acids): Facts, Fantasies & Failures. © 2014. 378 pages]

[Vitamin D: Benefits & False claims. © 2014. 206 pages]

[Chocolate & Red Wine Antioxidants (Polyphenols, Flavonoids & Resveratrol): Facts vs. Falsehoods. © 2015. 362 pages]

[Blueberry, Tomato & CoQ10 Antioxidants (Anthocyanins, Lycopene & Ubiquinone) Claims vs Facts. © 2015. 272 pages]

[Cancer Killing, Suppression and Protection: The Howes Answer to Cancer. © 2016. 486 pages]

The above books total about 8,000 pages of data and discussion.

MORTALITY

Antioxidant Supplements for Prevention of Mortality in Healthy Participants and Patients with Various Diseases.

Abstract
BACKGROUND:
Our systematic review has demonstrated that antioxidant supplements may increase mortality. We have now updated this review.
OBJECTIVES:
To assess the beneficial and harmful effects of antioxidant supplements for prevention of mortality in adults.
SEARCH METHODS:
We searched The Cochrane Library, MEDLINE, EMBASE, LILACS, the Science Citation Index Expanded, and Conference Proceedings Citation Index-Science to February 2011. We scanned bibliographies of relevant publications and asked pharmaceutical companies for additional trials.
SELECTION CRITERIA:
We included all primary and secondary prevention randomized clinical trials on antioxidant supplements (beta-carotene, vitamin A, vitamin C, vitamin E, and selenium) versus placebo or no intervention.
DATA COLLECTION AND ANALYSIS:
Three authors extracted data. Random-effects and fixed-effect model meta-analyses were conducted. Risk of bias was considered in order to minimize the risk of systematic errors. Trial sequential analyses were conducted to minimize the risk of random errors. Random-effects model meta-regression analyses were performed to assess sources of intertrial heterogeneity.
MAIN RESULTS:
Seventy-eight randomized trials with 296,707 participants were included. Fifty-six trials including 244,056 participants had low risk of bias. Twenty-six trials included 215,900 healthy participants. Fifty-two trials included 80,807 participants with various diseases in a stable phase. The mean age was 63 years (range 18 to 103 years). The mean proportion of women was 46%. Of the 78 trials, 46 used the parallel-group design, 30 the factorial design, and 2 the cross-over design. All antioxidants were administered orally, either alone or in combination with vitamins, minerals, or other interventions. The duration of supplementation varied from 28 days to 12 years (mean duration 3 years; median duration 2 years). Overall, the antioxidant supplements had no significant effect on mortality in a random-effects model meta-analysis (21,484 dead/183,749

(11.7%) versus 11,479 dead/112,958 (10.2%); 78 trials, relative risk (RR) 1.02, 95% confidence interval (CI) 0.98 to 1.05) but significantly increased mortality in a fixed-effect model (RR 1.03, 95% CI 1.01 to 1.05). Heterogeneity was low with an I(2)- of 12%. In meta-regression analysis, the risk of bias and type of antioxidant supplement were the only significant predictors of inter-trial heterogeneity. Meta-regression analysis did not find a significant difference in the estimated intervention effect in the primary prevention and the secondary prevention trials. In the 56 trials with a low risk of bias, the antioxidant supplements significantly increased mortality (18,833 dead/146,320 (12.9%) versus 10,320 dead/97,736 (10.6%); RR 1.04, 95% CI 1.01 to 1.07). This effect was confirmed by trial sequential analysis. Excluding factorial trials with potential confounding showed that 38 trials with low risk of bias demonstrated a significant increase in mortality (2822 dead/26,903 (10.5%) versus 2473 dead/26,052 (9.5%); RR 1.10, 95% CI 1.05 to 1.15). In trials with low risk of bias, beta-carotene (13,202 dead/96,003 (13.8%) versus 8556 dead/77,003 (11.1%); 26 trials, RR 1.05, 95% CI 1.01 to 1.09) and vitamin E (11,689 dead/97,523 (12.0%) versus 7561 dead/73,721 (10.3%); 46 trials, RR 1.03, 95% CI 1.00 to 1.05) significantly increased mortality, whereas vitamin A (3444 dead/24,596 (14.0%) versus 2249 dead/16,548 (13.6%); 12 trials, RR 1.07, 95% CI 0.97 to 1.18), vitamin C (3637 dead/36,659 (9.9%) versus 2717 dead/29,283 (9.3%); 29 trials, RR 1.02, 95% CI 0.98 to 1.07), and selenium (2670 dead/39,779 (6.7%) versus 1468 dead/22,961 (6.4%); 17 trials, RR 0.97, 95% CI 0.91 to 1.03) did not significantly affect mortality. In univariate meta-regression analysis, the dose of vitamin A was significantly associated with increased mortality (RR 1.0006, 95% CI 1.0002 to 1.001, P = 0.002).

AUTHORS' CONCLUSIONS:

We found no evidence to support antioxidant supplements for primary or secondary prevention. Beta-carotene and vitamin E seem to increase mortality, and so may higher doses of vitamin A. (Bjelakovic, Nikolova, Simonette and Gludd, 2012 Mar. 14)

Antioxidant supplements need to be considered as medicinal products and should undergo sufficient evaluation before marketing.

Antioxidants can increase your risk of dying. Taking antioxidant supplements containing beta-carotene, vitamin E, and vitamin A may be linked to an increased risk of death, according to a 2007 review and meta-analysis of 68 trials including a total of 232,606 participants. Although no increased mortality risk was associated with vitamin C supplementation, researchers didn't find any evidence that vitamin C supplements increased longevity either. (Bjlakovic pp. 842-57 et al, 2007)

https://www.researchgate.net/publication/258636670 Antioxidant supplements and mortality

Antioxidant supplements and mortality

Article · Literature Review · November 2013

Abstract

Oxidative damage to cells and tissues is considered by some to be involved in the aging process and in the development of chronic diseases in humans, including cancer and cardiovascular diseases, the leading causes of death in high-income countries. This has stimulated interest in the preventive potential of antioxidant supplements.

Today, more than one half of adults in high-income countries ingest antioxidant supplements foolishly hoping to improve their health, oppose unhealthy behaviors, and counteract the ravages of aging.

Older, more unreliable observational studies and some randomized clinical trials with high risks of systematic errors ('bias') have suggested that antioxidant supplements may improve health and prolong life.

A number of randomized clinical trials with adequate methodologies observed neutral or negative results of antioxidant supplements. **Recently completed large randomized clinical trials with low risks of bias and systematic reviews of randomized clinical trials taking systematic errors ('bias') and risks of random errors ('play of chance') into account have shown that antioxidant supplements do not seem to prevent cancer, cardiovascular diseases, or death. Even more, beta-carotene, vitamin A, and vitamin E may increase mortality.**

Some recent large observational studies now support these findings. According to recent dietary guidelines, there is no evidence to support the use of antioxidant supplements in the primary prevention of chronic diseases or mortality.

Antioxidant supplements do not possess preventive effects and may be harmful with unwanted consequences to our health, especially in well-nourished populations.

The optimal source of antioxidants seems to come from our diet, not from antioxidant supplements in pills or tablets.

Antioxidant supplements and mortality | Request PDF. Available from: https://www.researchgate.net/publication/258636670_Antioxidant_supplements_and_mortality [accessed Feb 14 2018].

---

https://www.medscape.org/viewarticle/552910

March 1, 2007 — **The largest analysis of data on antioxidant vitamins ever conducted has shown that beta-carotene, vitamin A, and vitamin E probably increase mortality**. Two other antioxidant substances — vitamin C and selenium — had no effect on mortality.

The meta-analysis of 68 randomized trials with a total of 232,606 participants, published in the February 28 issue of JAMA, was conducted by a group led by Goran Bjelakovic, MD, of the Copenhagen University Hospital in Denmark.

Coauthor, Christian Gluud, MD, of the Copenhagen University Hospital in Denmark, commented to heartwire : **"This is the most comprehensive collection of data on anti-oxidant vitamins ever conducted, and we have shown that on the whole that these agents have no benefit. Indeed, vitamin A, vitamin E and beta-carotene are associated with an increase in mortality at the doses studied**. Vitamin A and beta-carotene seem to have a dose-related effect, with mortality increasing as doses increase, whereas vitamin E does not appear to have a dose-related effect, with all doses associated with increased mortality."

Jury Still Out on Vitamin C and Selenium

Dr. Gluud added that the jury is still out on vitamin C and selenium. "**Vitamin C does not appear to be detrimental but it is not beneficial either, and all the trials of selenium together suggest a small benefit, but when only the well conducted trials are included, there appears to be neither benefit nor harm." "Our data show that anti-oxidant vitamins should not be taken in an effort to prevent illness.** People should instead eat a balanced diet and take regular exercise," he said.

In the JAMA article, **the authors note that many people are taking antioxidant supplements in the belief that they improve health and prevent diseases.**

Many primary or secondary prevention trials of antioxidant supplements have been conducted to prevent several diseases — mainly cardiovascular disease and cancer — but results have generally not been positive, with some trials showing increases in mortality.

To find out more, they conducted the current systematic review to analyze the effects of antioxidant supplements on all-cause mortality of adults included in primary and secondary prevention trials. They included all primary and secondary prevention published trials in adults randomized to receive beta-carotene, vitamin A, vitamin C, vitamin E, or selenium vs placebo or no intervention.

Results showed that when all trials of antioxidant supplements were pooled together, there was no significant effect on mortality, but when the 47 trials said to have a low risk of bias (in a total of 180,938 participants) were analyzed alone, the antioxidant supplements as a whole significantly increased mortality, and beta-carotene, vitamin A, and vitamin E were all associated with increased mortality when given alone or in combination. Vitamin C and selenium had no significant effect on mortality.

Source: JAMA. 2007; 297:842-857.

The researchers note that more than two thirds of the included trials fell into the category of low-bias risk trials, which they say highlights the validity of their results. "Antioxidant supplements not only seem to be one of the most researched topics in the world, they also seem to be one of the most adequately researched clinical questions," they say.

They point out that a large number of unpublished trials on supplements may exist, but as unpublished trials are more likely to have been either neutral or negative than to have shown beneficial effects, this suggests their estimate of a 5% increase in mortality is likely to be conservative.

Substantial Public Health Consequences

Noting that 10% to 20% of the adult population (80 - 160 million people) in North America and Europe may consume these supplements, Bjelakovic and colleagues say the public health consequences may be substantial.

Speculating on possible mechanisms, they point out that although oxidative stress has a hypothesized role in the pathogenesis of many chronic diseases, it may be the consequence of pathologic conditions, and that by eliminating free radicals, some essential defensive mechanisms may be affected.

In an interview with Heartwire, Dr. Gluud also suggested that the antioxidant vitamins could actually also have prooxidant effects. "We don't know exactly how they are doing harm but rather than preventing cardiovascular disease and cancer, they actually seem to be accelerating these conditions."

Lessons Learned

Dr. Gluud said these observations were "a huge disappointment," but added that at least it has been discovered. "**We must see the positives in this. The question has been thoroughly addressed and we now know the answer — these agents are harmful**. The companies selling these anti-oxidant vitamins have been able to dodge the issue for a long time, saying that any negative data has not been comprehensive. They cannot do this any longer. There are lessons to be learnt here. For example, the importance of conducting trials with these agents and publishing the results."

Dr. Gluud added that food supplements should be regulated in the same way as medical products. "The governments of the world now have the responsibility to inform people of these results. They have been too slow in the past in requesting that health supplements are properly evaluated and allowing these products to be added to foods. People have been buying these supplements and foods advertised as having these supplements added under the impression that they are good for them, **when in actual fact they are harmful.** Any potential health supplements should not be allowed to be added to foods unless they have been shown to be beneficial, or at least proven not to be harmful."

(Bjlakovic pp. 842-57 et al, 2007)

SUICIDE

Relative to antioxidants, the average lay person is unknowingly committing suicide.

They are unintentionally committing suicide out of ignorance.

Victims have been hoodwinked into believing antioxidants are "super foods" and the cure to all human ailments or diseases, including aging.

The scientific watering hole was poisoned by Dr. Denham Harman's erroneous free radical theory and has dominated the thinking of scientists for well over half a century. The free radical theory is an outdated, debunked bunch of hooey.

Antioxidant suicide victims have been misled by advertisers, marketeers, supplement manufacturers and uninformed so-called medical scientists but in the end, excessive ingestion of antioxidants is still plain ole "suicide."

Harman's Lamentable Free Radical Theory

Denham Harman proposed in 1956 the "free radical theory," speculating that damage to aerobic organisms occurs due to harmful free radical production of oxidative products.

Subsequently, these alleged damaging derivatives of oxygen, which were termed either oxygen free radicals or "reactive oxygen species (ROS)," were defined as being deleterious and harmful. For greater accuracy, I will use the term "electronically modified oxygen derivatives (EMODs)" to replace the less accurate term of reactive oxygen species.

Unfortunately, the overly optimistic predictions based on the free radical theory have repeatedly failed to scientifically support the free radical theory.

The Harvard School of Public Health web site summed it up this way, "The evidence accumulated thus far on antioxidant vitamins isn't promising. Randomized trials of vitamin C, vitamin E, and beta-carotene haven't revealed much in the way of protection from heart disease, cancer, or aging-related eye diseases (website accessed 2-09-06).

The free radical theory erroneously stated that diseases and the aging process resulted from the "stochastic" accumulation of oxidative damage purportedly caused by EMODs, from sources such as the environment and from normal by-products of cellular metabolism.

Contrary to the free radical theory of aging, which argues that EMODs are uncontrolled, EMODs are under strict metabolic control. There is a compartmentalization of oxidative events, which strongly suggests EMODs' crucial role influencing and modulating physiological stimuli, signaling mechanisms, and functional homeostasis.

Tests of effect of vitamin E and other antioxidant vitamins or their combinations on clinical manifestations of cardiovascular disease, cancer and diabetes, have consistently shown that

commonly used antioxidant vitamin regimens (vitamins E, C, beta carotene, or a combination thereof) do not significantly reduce overall cardiovascular events, diabetes or cancer in studies such as HOPE, GISSI, ATBC, Hennekens study, Omenn's study, Brown's study, MRC/BHF, Vivekananthan's meta-study, Miller's meta-study (Yusuf et al, 2000) (GISSI, 1999) (Heinonen et al, 1994) (Hennekens et al, 1996) (Omenn et al, NEJM, 1996) (Brown et al, 2001) (Collins et al, 2002) (Vivekanathan et al, 2003) (Miller et al, 2005).

Antioxidants actually appear to cause harm and in some studies they may increase overall mortality, which is discussed in a 2005 nutrition and supplement review in JAMA.

Inhibition of apoptosis by antioxidants may explain why, in several studies in heavy smokers, vitamin E and β-carotene enhanced carcinogenesis in the lung. Increased formation of EMODs also accompanies apoptosis induced by most, if not all, other stimuli, and free radical scavengers nearly always delay such apoptosis (Hampton et al, 1998) (Jacobson, Raff, 1995) (Huang et al, 2000) (Neil, Kay, 2006) (Chandra et al, 2003) (Shanafelt et al, 2005) (Mow et al, 2002) (Ling et al, 2003) (Hileman et al, 2004) (Choi, Singh, 2005) (Singh et al, 2005).

Davies has shown that cellular division or cell death is EMOD concentration dependent, when utilizing the EMOD, H_2O_2. Cellular responses go from proliferation, to arrest, to apoptosis (Davies, 1999).

Great thoughts are a relative rarity;
whereas,
great mistakes are commonplace.
R.M. Howes, M.D., Ph.D.
8-19-04

It requires 3 times more radiation to kill cancer cells in the absence of O_2 than it does in the presence of normal levels of O_2. Hypoxia-associated resistance to photon radiotherapy is multifactorial. The presence of molecular oxygen increased DNA damage through the formation of oxygen free radicals, which occurs primarily after the interaction of radiation with intracellular water.

I have presented approximately 106 studies showing EMODs are responsible for induction of apoptosis.

I have presented approximately 50 studies showing that antioxidants can block EMOD induced apoptosis.

In this chapter there are 24 studies showing that antioxidants increased the risk of cancer and 2 showing they increased the mortality in cancer patients (26 total).

Antioxidants Increased the Risk of the Following Cancers:
- lung cancer, lung cancer, lung cancer, lung cancer, lung cancer, lung cancer among smokers, lung cancer, (7 studies)
- advanced prostate cancer, aggressive or fatal prostate cancer, prostate cancer death rates, fatal prostate cancer, prostate cancer, (5 studies)
- doubled the risk of adenoma recurrence,
- squamous cell carcinoma and total nonmelanoma skin cancer, squamous cell cancer (2 studies)
- melanoma
- higher rate of second primary cancers (head and neck), local recurrence of the head and neck tumor tended to be higher, (2 studies)
- esophageal cancer, (2 studies)
- oral pre-malignant lesions,
- stomach cancer
- urinary bladder cancer
- breast cancer

SUMMARY: Lung Cancer, Breast Cancer, Prostate Cancer, Stomach Cancer, Esophageal Cancer, Head and Neck Cancer, Bladder Cancer, Melanoma, Adenomas, Oral Premalignant Lesions

Antioxidants increased the mortality in the following cancer patients (2 studies):

- increase overall mortality in gastrointestinal cancer patients,

- esophageal cancer deaths increased 14% in patients, (2 studies)

A state of health is only maintained
by a continuously vigilant oxidative defensive system,
without which we quickly succumb to
bugs, drugs and thugs called cancer cells.
Debilities and anomalies of our somatic condition are
pounced upon by pathogens and mutagens
but they are quickly
"rebuffed by radicals."
Oxidative "radical outbursts" of EMODs
are proven tireless workers.
So it is in the world of
protective homeostasis.
R.M. Howes, M.D., Ph.D.
6-11-06

One hundred eighteen (118) studies showing antioxidant ineffectiveness in cancer treatment and thirty-nine (39) studies showing harmful effects (taken from book Two, *Death in small doses?* and from *Antioxidant Overkill*)

Some of the following was excerpted or modified from Wikipedia: https://en.wikipedia.org/wiki/Suicide

Suicide and Self-Harm

The word suicide is from the Latin suicidium, which means "the killing of oneself".

Suicide is the act of intentionally causing one's own death. Suicide prevention efforts include limiting access to methods of suicide, such as firearms, drugs, and poisons, and may I add ingestion of excessive amounts of antioxidants. The best way to minimize the risk of suicide is to know the risk factors and I am therefore presenting you with the considerable data showing increased mortality rates with the ingestion of antioxidants. Hence, antioxidant suicide is a reality.

In ancient Athens, a person who committed suicide without the approval of the state was denied the honors of a normal burial. The person would be buried alone, on the outskirts of the city, without a headstone or marker. However, it was deemed to be an acceptable method to deal with military defeat. In Ancient Rome, while suicide was initially permitted, it was later deemed a crime against the state due to its economic costs. Aristotle condemned all forms of suicide while Plato was ambivalent. In Rome some reasons for suicide included volunteering death in a gladiator combat, guilt over murdering someone, to save the life of another, as a result of mourning, from shame from being raped, and as an escape from intolerable situations like physical suffering, military defeat, or criminal pursuit.

Suicide came to be regarded as a sin in Christian Europe and was condemned at the Council of Arles as the work of the Devil. In the Middle Ages, the Church had drawn-out discussions as to when the desire for martyrdom was suicidal, as in the case of martyrs of Córdoba. Despite these disputes and occasional official rulings, Catholic doctrine was not entirely settled on the subject of suicide until the later 17th century. A criminal ordinance issued by Louis XIV of France in 1670 was extremely severe, even for the times: the dead person's body was drawn through the streets, face down, and then hung or thrown on a garbage heap. Additionally, all of the person's property was confiscated.

Attitudes towards suicide slowly began to shift during the Renaissance. John Donne's work Biathanatos, contained one of the first modern defenses of suicide, bringing proof from the conduct of Biblical figures, such as Jesus, Samson and Saul, and presenting arguments on grounds of reason and nature to sanction suicide in certain circumstances.

The secularization of society that began during The Enlightenment questioned traditional religious attitudes toward suicide and brought a more modern perspective to the issue. David Hume denied that suicide was a crime as it affected no one and was potentially to the advantage of the individual. In his 1777 Essays on Suicide and the Immortality of the Soul he rhetorically asked, "Why should I prolong a miserable existence, because of some frivolous advantage which the public may perhaps receive from me?" A shift in public opinion at large can also be discerned; The Times in 1786 initiated a spirited debate on the motion "Is suicide an act of courage?".

By the 19th-century, the act of suicide had shifted from being viewed as caused by sin to being caused by insanity in Europe. Although suicide remained illegal during this period, it increasingly became the target of satirical comments, such as the Gilbert and Sullivan comic opera The Mikado that satirized the idea of executing someone who had already killed himself.

By 1879, English law began to distinguish between suicide and homicide, although suicide still resulted in forfeiture of estate. In 1882, the deceased were permitted daylight burial in England and by the middle of the 20th century, suicide had become legal in much of the western world. The term suicide first emerged shortly before 1700 to replace expressions on self-death which were often characterized as a form of self-murder in the West.

In most Western countries, suicide is no longer a crime. It was, however, in most Western European countries from the Middle Ages until at least the 1800s. It remains a criminal offense in most Muslim-majority nations.

In the United States, suicide is not illegal but may be associated with penalties for those who attempt it. Physician-assisted suicide is legal in the state of Washington for people with terminal diseases. In Oregon, people with terminal diseases may request medications to help end their life.

Canadians who have attempted suicide may be barred from entering the US. US laws allow border guards to deny access to people who have a mental illness, including those with previous suicide attempts.

Islamic religious views are against suicide. The Qu'ran forbids it by stating "do not kill or destroy yourself". The hadiths also state individual suicide to be unlawful and a sin. Stigma is often associated with suicide in Islamic countries.

In Hinduism, suicide is generally frowned upon and is considered equally sinful as murdering another in contemporary Hindu society. Hindu Scriptures state that one who dies by suicide will become part of the spirit world, wandering earth until the time one would have otherwise died, had one not taken one's own life. However, Hinduism accepts a man's right to end one's life through the non-violent practice of fasting to death, termed Prayopavesa. But, Prayopavesa is strictly restricted to people who have no desire or ambition left, and no responsibilities remaining in this life. Jainism has a similar practice named Santhara. Sati, or self-immolation by widows, was prevalent in Hindu society during the Middle Ages.

Some landmarks have become known for high levels of suicide attempts. These include San Francisco's Golden Gate Bridge, Japan's Aokigahara Forest, England's Beachy Head and Toronto's Bloor Street Viaduct.

As of 2010, the Golden Gate Bridge has had more than 1,300 die by suicide by jumping since its construction in 1937. Many locations where suicide is common have constructed barriers to prevent it; this includes the Luminous Veil in Toronto, the Eiffel Tower in Paris and Empire State Building in New York City. They appear to be generally effective.

An example of mass suicide is the 1978 Jonestown killings/suicide in which 909 members of the Peoples Temple, an American religious group led by Jim Jones, ended their lives by drinking grape Flavor Aid laced with cyanide and various prescription drugs. Thousands of Japanese civilians took their own lives in the last days of the Battle of Saipan in 1944, some jumping from "Suicide Cliff" and "Banzai Cliff".

The 1981 hunger strikes, led by Bobby Sands, resulted in 10 deaths. The cause of death was recorded by the coroner as "starvation, self-imposed" rather than suicide; this was modified to simply "starvation" on the death certificates after protest from the dead strikers' families. During World War II, Erwin Rommel was found to have foreknowledge of the July 20 Plot on Hitler's life; he was threatened with public trial, execution and reprisals on his family unless he took his own life.

Common methods include hanging, pesticide poisoning, and firearms. Suicide resulted in 828,000 deaths globally in 2015 (up from 712,000 deaths in 1990). This makes it the 10th leading cause of death worldwide. Logically, deaths from antioxidant ingestion should be added to this total but it is more insidious than these other causes.

Views on suicide have been influenced by broad existential themes such as religion, honor, and the meaning of life. The Abrahamic religions traditionally consider suicide an offense towards God due to the belief in the sanctity of life.

During the samurai era in Japan, a form of suicide known as seppuku (harakiri) was respected as a means of making up for failure or as a form of protest. Sati, a practice outlawed by the British Raj, expected the Indian widow to kill herself on her husband's funeral fire, either willingly or under pressure from her family and society.

Suicide and attempted suicide, while previously illegal, are no longer so in most Western countries. It remains a criminal offense in many countries. In the 20th and 21st centuries, suicide has been used on rare occasions as a form of protest, and kamikaze and suicide bombings have been used as a military or terrorist tactic.

Attempted suicide or non-fatal suicidal behavior is self-injury with the desire to end one's life that does not result in death. Assisted suicide is when one individual helps another bring about their own death indirectly via providing either advice or the means to the end. This is in contrast to euthanasia, where another person takes a more active role in bringing about a person's death. Suicidal ideation is thoughts of ending one's life but not taking any active efforts to do so.

In a murder-suicide (or homicide-suicide), the individual aims at taking the life of others at the same time.

Genetics appears to account for between 38% and 55% of suicidal behaviors.

Eating disorders are another high-risk condition.

In approximately 80% of completed suicides, the individual has seen a physician within the year before their death, including 45% within the prior month. Approximately 25–40% of those who completed suicide had contact with mental health services in the prior year.

Substance abuse is the second most common risk factor for suicide after major depression and bipolar disorder.

Smoking cigarettes is associated with risk of suicide. There is little evidence as to why this association exists; however, it has been hypothesized that those who are predisposed to smoking are also predisposed to suicide, that smoking causes health problems which subsequently make people want to end their life, and that smoking affects brain chemistry causing a propensity for suicide. Cannabis however does not appear to independently increase the risk.

In Japan, health problems are listed as the primary justification for suicide.

This trigger of suicide contagion or copycat suicide is known as the Werther effect, named after the protagonist in Goethe's The Sorrows of Young Werther who killed himself and then was emulated by many admirers of the book.

The opposite of the Werther effect is the proposed Papageno effect, in which coverage of effective coping mechanisms may have a protective effect. The term is based upon a character in Mozart's opera The Magic Flute, who (fearing the loss of a loved one) had planned to kill himself until his friends helped him out. When media follows recommended reporting guidelines the risk of suicides can be decreased. Getting buy-in from industry, however, can be difficult, especially in the long term.

Rational suicide is the reasoned taking of one's own life, although others consider suicide as never rational. The act of taking one's life for the benefit of others is known as altruistic suicide. An example of this is an elder ending his or her life to leave greater amounts of food for the younger people in the community. Suicide in some Inuit cultures has been seen as an act of respect, courage, or wisdom.

Worldwide, 30% of suicides are estimated to occur from pesticide poisoning, most of which occur in the developing world. The use of this method varies markedly from 4% in Europe to more than 50% in the Pacific region. It is also common in Latin America due to easy access within the farming populations. In many countries, drug overdoses account for approximately 60% of suicides among women and 30% among men. Many are unplanned and occur during an acute period of ambivalence.

Together, hanging and poisoning comprised about 40% of U.S. suicides (as of 2005).

There is no known unifying underlying pathophysiology for either suicide or depression.

Suicide prevention is a term used for the collective efforts to reduce the incidence of suicide through preventative measures. **Reducing access to certain methods, such as firearms or toxins can reduce risk**.

Some have proposed reducing access to alcohol as a preventative strategy (such as reducing the number of bars).

Approximately 0.5% to 1.4% of people die by suicide, a mortality rate of 11.6 per 100,000 persons per year. Suicide resulted in 842,000 deaths in 2013 up from 712,000 deaths in 1990.

Extraordinary theories necessitate extraordinary proof.
I have amassed epic support for my innovative prooxidant theories to conquer cancer, heart disease, HIV/AIDS and malaria, the biggest global killers of mankind.

R. M. Howes, M.D., Ph.D.

11-7-10

The following papers illustrate the state of the current thinking regarding Harman's free radical theory and the free radical theory of aging. They also show the debunked role of antioxidants in fighting so-called free radical diseases, including cancer.

**The story of antioxidants,
as they relate to disease
prevention, cure, and antiaging
is the story of ….
FAILURE!**
R. M. Howes, M.D., Ph.D.
6-2-11

The antioxidant supplement market is quickly becoming a gargantuan crime scene. Fraudulent claims, unsupported hype and outright lies are the languages of choice for their $28 billion industry. Reportedly, the dietary supplement industry income now exceeds $53 billion annually.

Antioxidants are frequently used and promoted in so-called anti-aging programs. The Business Communications Company Research firm said the U.S. market for anti-aging products is worth $45.5 billion and is growing nearly 10 percent yearly.

Yet, no single agent has been shown to truly reverse aging or increase longevity in humans (Kamel et al, 2006). Scams are everywhere.

Reactive Oxygen Species and the Free Radical Theory of Aging

Abstract

The traditional view in the field of free radical biology is that free radicals and reactive oxygen species (ROS) are toxic, mostly owing to direct damage of sensitive and biologically significant targets and are thus a major cause of oxidative stress; that complex enzymatic and nonenzymatic systems act in concert to counteract this toxicity; and that a major protective role is played by the phenomenon of adaptation. Another part of the traditional view is that the process of aging is at least partly due to accumulated damage done by these harmful species. However, **recent workers in this and in related fields are exploring the view that superoxide radical and reactive oxygen species exert beneficial effects. Thus, such ROS are viewed as involved in cellular regulation by acting as (redox) signals, and their harmful effects are seen mostly as a result of compromised signaling, rather than due to direct damage to sensitive targets.** According to some followers of this view, **ROS such as hydrogen peroxide and superoxide are not just causative agents of aging but may also be agents that increase the life span by acting, for example, as prosurvival signals**. (Liochev Sl, 2013)

The Role of Reactive Oxygen Species and Mitochondria in Aging

Abstract

To date, multiple theories have been proposed to explain the causes of aging. One of them is the free radical theory which postulates that reactive oxygen species (ROS), extremely reactive chemical molecules, are the major cause of the aging process. These free radicals are mainly produced by the mitochondrial respiratory chain as a result of electron transport and the reduction of the oxygen molecule. Toxic effects of ROS on cellular components lead to accumulation of oxidative damage which causes cellular dysfunction with age. The free radical theory has been one of the most popular theories of aging for many years. Scientific research on different model organisms aiming to verify the theory has produced abundant data, supporting the theory or, on the contrary, suggesting strong evidence against it. **At present, the free radical theory of aging is no longer considered to be true.** (Piotrowska A., Bartnik E, 2014)

A Midlife Crisis for the Mitochondrial Free Radical Theory of Aging

(Stuart JA, et al., 2014)

Abstract

Since its inception more than four decades ago, the Mitochondrial Free Radical Theory of Aging (MFRTA) has served as a touchstone for research into the biology of aging. The MFRTA suggests that oxidative damage to cellular macromolecules caused by reactive oxygen species (ROS) originating from mitochondria accumulates in cells over an animal's lifespan and eventually leads to the dysfunction and failure that characterizes aging. A central prediction of the theory is that the ability to ameliorate or slow this process should be associated with a slowed rate of aging and thus increased lifespan. A vast pool of data bearing on this idea has now been published.

ROS production, ROS neutralization and macromolecule repair have all been extensively studied in the context of longevity. We review experimental evidence from comparisons between naturally long- or short-lived animal species, from calorie restricted animals, and from genetically modified animals and weigh the strength of results supporting the MFRTA. Viewed as a whole, the data accumulated from these studies have too often failed to support the theory. **Excellent, well controlled studies from the past decade in particular have isolated ROS as an experimental variable and have shown no relationship between its production or neutralization and aging or longevity.** Instead, a role for mitochondrial ROS as intracellular messengers involved in the regulation of some basic cellular processes, such as proliferation, differentiation and death, has emerged.

Antioxidants

The second prediction arising from the MFRTA is that greater longevity should be associated with a greater capacity to neutralize mitochondrial ROS. Within the mitochondrial matrix Mn O_2^- dismutase (MnSOD) catalyzes the conversion of O_2^- to H_2O_2 in a diffusion-rate limited reaction.

The inner membrane is impermeable to $O_2^{\cdot-}$ due to this molecule's negative charge, and as the sole $O_2^{\cdot-}$ dismutase in the matrix MnSOD therefore acts as a primary regulator of $O_2^{\cdot-}$ concentration in this compartment and is important in controlling concentrations of ROS generated from $O_2^{\cdot-}$ produced by mitochondria. $O_2^{\cdot-}$ produced on the IMS side of the electron transport chain is converted to H_2O_2 by CuZnSOD, a primarily cytosolic antioxidant enzyme that has also been associated with the IMS. In rat liver, nearly 3% of the total cellular concentration of this enzyme is localized to the mitochondrial IMS. **H_2O_2 generated by $O_2^{\cdot-}$ dismutation in the matrix may go on to be further detoxified to water within mitochondria by glutathione peroxidase, peroxiredoxin 3 and 5, and thioredoxin pathways or, in heart mitochondria, catalase. Mitochondrial H_2O_2 that is not intercepted by antioxidant enzymes in the matrix can diffuse into the cytosol, where it may be detoxified by cytosolic isozymes in the glutathione and thioredoxin pathways, or by the enzyme catalase**.

Measurements of the two major $O_2^{\cdot-}$ dismutases of the mitochondrial (MnSOD) and cytosolic (CuZnSOD) compartments and several enzymes involved in H_2O_2 neutralization (catalase and also the cycle of glutathione oxidation involving glutathione peroxidase and glutathione reductase) have been made in many of the same experimental models used for assessing mitochondrial ROS production.

In a comparison of the naked mole rat and laboratory mouse, activities of MnSOD (not corrected for mitochondrial abundance) and CuZnSOD measured in liver at mid-age are significantly higher in the naked mole rat. In contrast, catalase activities are not different and glutathione peroxidase activities are an order of magnitude lower in the naked mole rat liver.

Page *et al.* measured all five of the antioxidant enzymes listed above in brain, heart and liver tissues of 14 species of endotherm vertebrates. (Page *et al.*, 2010)

Of 15 tested correlations (five enzymes x three tissues), only two were positive and statistically significant. **These were MnSOD and catalase in the brain, which were higher in longer-lived species**, even after correction for body mass and phylogenetic effects. **Similar measurements of two other antioxidant enzymes, glutaredoxin and thioredoxin reductase, also failed to reveal significant positive correlations with lifespan in any of these three tissues. Thus, of 21 tested associations of six antioxidant enzymes only 2 showed the hypothesized positive correlations with lifespan**. Since measurements made in whole tissue homogenates provide little insight into ROS neutralization within mitochondria, we measured glutathione peroxidase and glutathione reductase activities in brain mitochondria isolated from eight species of vertebrate endotherm. **This analysis also failed to show a relationship between with MLSP, and therefore failed to support the second hypothesis relating to the MFRTA, that is, that the cellular capacity to neutralize ROS should be greater in longer-lived organisms.**

Walsh *et al.* recently summarized the results of several decades of studies examining antioxidant enzyme levels/activities (superoxide dismutases, catalase, glutathione metabolizing enzymes) in the context of caloric restriction. Similar to their findings with respect to mitochondrial ROS production, **these authors show the absence of a consistent up-regulation of antioxidant enzymes concomitant with caloric restriction in mice and rats.** (Walsh *et al.*, 2014)

A number of mammalian lifespan studies have been conducted utilizing transgenic or knockout laboratory mouse models to increase or decrease gene expression of mitochondrial and other key intracellular antioxidant enzymes. The results of such studies have been the in-depth focus of other review papers and, therefore, will not be reviewed in detail here. **Overwhelmingly, the conclusions from these studies has been that, although the expected increases and decreases in tissue oxidative damage biomarkers are usually observed in antioxidant enzyme gene under-expressing and overexpressing individuals, respectively, there are seldom corresponding effects on longevity. Thus, the results of experiments using this approach have most often yielded results that are inconsistent with the MFRTA.**

Small molecule antioxidants have been promoted extensively to the general public as anti-aging and pro-longevity supplements. The evidentiary underpinnings of this are rooted in part in the observations of pro-health effects of various plant-based foods with antioxidant constituents. Hundreds of experiments have now been completed to examine the putative anti-aging effects of vitamin E (tocopherols and tocotrienols) in a diverse range of species from protists to mammals, and the results of these experiments have been reviewed recently.

Vitamin E has variously been shown to have no effect, a positive effect and even a negative effect on aging/lifespan. Certainly, no clear picture of an anti-aging activity emerges in the hundreds of studies that have been conducted. This includes human studies, some of which have been terminated prematurely due to adverse outcomes. A similar lack of consensus has emerged with respect to the anti-aging effects of a number of other vitamin antioxidant supplements, after many hundreds of experimental studies and clinical trials.

A variety of plant-based molecules, including polyphenolic stilbenes, such as resveratrol, have more recently been put forth as anti-aging elixirs due in part to their antioxidant activities. Although early results seemed to suggest pro-longevity properties for resveratrol, the dozens of experiments instigated by these findings failed to confirm any general positive effects. While there is some evidence for increased lifespan in *C. elegans*, it is lacking in most other species.

The National Institutes of Health's Aging Intervention Testing Study (http://www.nia.nih.gov/research/dab/interventions-testing-program-itp/compounds-testing) has investigated the pro-longevity properties of a number of small molecule antioxidants, including vitamin E and resveratrol, in mice and reported no beneficial effects on lifespan.

Based on the results discussed above, the evidence for an association between small molecule antioxidant supplementation and slowed aging and/or increased longevity is insufficient to support the MFRTA.

However, it is important to note that none of these tested molecules is specifically targeted to mitochondria, so the extent to which they access the organelle in any tissue or cell is likely highly variable. To address this potential limitation, some investigators have developed antioxidants conjugated to positively charged, membrane-permeant moieties that target them specifically to mitochondria. Perhaps the best studied example is the mitochondria targeted ubiquinone (MitoQ). **The anti-aging properties of MitoQ have been tested in *D. melanogaster*, where it failed to extend lifespan.** While we await further evidence of the ability of MitoQ, or other mitochondria-targeted antioxidants, to slow the rate of aging, **at this time there is no compelling evidence that reducing the rate of mitochondrial ROS production will slow aging or increase lifespan. Therefore, this line of investigation has failed to offer clear support for the MFRTA.**

Conclusions

The MFRTA has stimulated an enormous amount of research into the role of mitochondrial ROS production and oxidative stress in aging and longevity. However, **as it enters its fifth decade, it seems to be having something of a mid-life crisis. Virtually all attempts to control mitochondrial ROS production or neutralization have yielded unexpected and even occasionally unwanted effects on aging and lifespan.** And it seems that those organisms that have (at least partly) solved the riddle of longevity have not done so by addressing the 'ROS problem'.

Thus, **the MFRTA has as yet failed to offer a sufficient explanation of organismal aging as a phenomenon.**

Methodological limitations may be invoked to explain the inability to detect the predicted relationships among mitochondrial ROS production, neutralization, and macromolecule damage and repair in any specific context. However, it is more difficult to advance this argument in the context of the many quite different approaches that have been taken and **failed to consistently validate the predictions.**

Whether considering the evolution of longevity by natural selection of specific traits, the extension of lifespan by caloric restriction, the ability of transgenes, gene knockouts or small molecule antioxidants to alter lifespan, the overall conclusion has been drifting toward 'no consistent relationship between mitochondrial ROS and longevity'.

Nonetheless, investigation of the MFRTA has contributed to the increasing depth of our understanding of ROS activities in animal cells. ROS are recognized to impinge upon signaling pathways regulating all of the fundamental aspects of cell biology: the cell cycle, proliferation and differentiation, and life and death.

There is No Evidence that Mitochondria are the Main Source of Reactive Oxygen Species in Mammalian Cells.

(Brown GC, Borutaite V., 2012)

Abstract
It is often assumed that mitochondria are the main source of reactive oxygen species (ROS) in mammalian cells, but there is no convincing experimental evidence for this in the

literature. What evidence there is suggests mitochondria are a significant source for ROS, which may have physiological and pathological effects. But quantitatively, endoplasmic reticulum and peroxisomes have a greater capacity to produce ROS than mitochondria, at least in liver. Mitochondria can rapidly degrade ROS and thus are potential sinks for ROS, but whether mitochondria act as net sources or sinks within cells in particular conditions is unknown.

New Challenges to Study Heterogeneity in Cancer Redox Metabolism

(Benfeitas R, Uhlen M, Nielsen J, Mardinoglu A., 2017)

Abstract

Considering that ROS levels are significantly increased in cancer cells due to mitochondrial dysfunction, ROS metabolism has been targeted for the development of efficient treatment strategies, and antioxidants are used as potential chemotherapeutic drugs. However, **initial ROS-focused clinical trials in which antioxidants were supplemented to patients provided inconsistent results, i.e., improved treatment or increased malignancy**.

In turn, H_2O_2 has emerged as the ROS that displays the best signaling properties. Its high stability and selective reactions with cellular compounds permit diffusion over distances of several micrometers and enable cell membrane crossing, which is also facilitated through specific channels. H_2O_2 reacts with cellular thiols, including those contained in low molecular weight compounds, such as glutathione, and protein thiols, such as peroxiredoxins and thioredoxins. These reactions convert an oxidizing equivalent into a redox signal, which may be transduced from protein to protein via thiol disulfide exchange or between glutathione and proteins, forming mixed disulfides. Together with intracellular thiols, H_2O_2 regulates the redox state and activity of several target proteins and has pivotal importance in both physiological and pathological conditions.

The role of ROS in cell physiology is highly dependent on their levels. Under physiological levels, ROS regulate a number of signaling processes by reacting with proteins, genes, and transcription factors. ROS control adaptation to hypoxia, regulation of differentiation, immunity, and longevity. However, **the accumulation of ROS beyond physiological levels promotes cell proliferation, angiogenesis, and even apoptosis** and ROS also control cell-cycle progression.

High antioxidant activities enable fast ROS-driven proliferation and metastasizing, but the increased oxidative stress is insufficient to lead to apoptosis. Higher antioxidant expression is also associated with the radioresistance observed in certain cancer stem cell populations. In turn, recent antioxidant-targeted therapeutic strategies have shifted their focus in the opposite direction, exploiting ROS toxicity as a means leading to cancer cell apoptosis. **These drugs often mimic ROS-generating enzymes (e.g., NADPH oxidases and superoxide dismutases), inhibit antioxidant enzymes (e.g., catalase), deplete thiol pools, such as GSH, or shift redox buffer ratios (e.g., GSSG/2 GSH), thereby promoting a more oxidizing intracellular state and cell apoptosis.** (Gupta et al., 2012).

The influence of ROS on different biochemical levels makes it necessary to seek an integrative analysis of these systems at the genomic, proteomic, and metabolomic levels. Approaches that are able to encompass these levels and integrate the crosstalk between antioxidant, redox, energetic, and central metabolism are able to capture and understand these complex responses.

Systems biology approaches may be used to analyze omics data and understand the roles of each redox system in cancer.

Antioxidants: Pro or Anti?

Life, and aging, used to be so simple. Findings that older organisms, from helminths to mice to men, tend to generate more free radicals in their cells (specifically, in their mitochondria), and show more signs of oxidative damage, gave rise to the Mitochondrial Free Radical Theory of Aging. This iconic theory, otherwise known as MFRTA, has been around in various versions since the mid-sixties (Harman '66); and it has been used to sell tons of antioxidant supplements. Unfortunately, like so many other 60's icons, MFRTA has not worn well, and in the last few years we have seen a steady stream of experimental findings that have left it, by now, fatally wounded.

The scientific method demands that we change our beliefs or theories to fit the factual data. I believe that this applies directly to the Free Radi-Crap theory. Again, I say to you, "The free radical theory has fallen and so has the mitochondrial free radical theory of aging."

Antioxidants Can Even Kill You

First, please remember the work of Drisko et al with IV mega-doses of vitamin C which resulted in 2 deaths and other adverse effects. Then check out the deaths from clioquinol and of 38 infant deaths from vitamin E. (Padayatty et al, 2010).

Deaths in Japan: Clioquinol

In 1955 a mysterious disease, **resembling polio**, appeared in Japan and had symptoms of a **combination of diarrhea, internal bleeding and various signs of nerve degeneration. By 1959 the disease had become an epidemic.** The illness appeared to be **contagious**, but patients did not display the symptoms typically associated with infections. **By 1964 the epidemic had worsened, and patients were exhibiting blindness, with some patients dying.**

During the 1960s thousands of Japanese users of the anti-diarrhea drug clioquinol were left crippled, blinded, or otherwise disabled (sometimes leading to death) by a nerve disease known as subacute myelo-optic neuropathy (SMON). Dozens of elderly women, and some men in their thirties, began filling the nearby hospitals, totaling almost 3 percent of the local population by 1971.

In May of 1964, the disease was given a formal name: **"Sub-acute Myelo-Optic Neuropathy" (SMON) and by 1971, the number of people hospitalized in the Okayama Province accounted for about three per cent of the province's population. SMON victims had received treatment for diarrhea with a number of drugs.** Upon investigation, **these different drugs turned out not to be different at all; they were all made of a substance called Clioquinol, an antioxidant, but marketed under different brand names and freely available.**

The antioxidant Clioquinol, a Ciba-Geigy product, was erroneously considered to be perfectly safe, because its effects would be confined to the digestive tract where it was supposed to destroy germs associated with diarrhea without being absorbed into the bloodstream. But the evidence was irrefutable and the SMON epidemic lasted until just after the government finally banned the antioxidant drug in September 1970. But Ciba-Geigy nevertheless continued selling the drug worldwide.

The Japanese government recognized about 11,000 SMON victims, 4,700 of whom had filed damage claims as of 1979. (Chapmann, 1979).

Curiously, overmedication is more common in Japan than elsewhere because doctors receive payment from the government health insurance for every drug they prescribe.

Clioquinol (5-chloro-7-iodo-8-hydroxyquinoline; CQ) belongs to the quinoline class of compounds and is structurally similar to 5,7-DiCl-8-OHQ. This class of compound possesses an established toxicology profile with the US Pharmacopoeia. During the 1950s to the 1970s, CQ was used as an antibiotic; however, it was withdrawn due to association with subacute myelo-optic neuropathy possibly due to overdose and/or a reversible vitamin B_{12} deficiency. Recently, interest in CQ has reemerged due to studies involving its use, in combination with B_{12}, for treatment of Alzheimer's disease. It has a very controversial history.

Clioquinol (5-chloro-7-iodo-8-quinolinol) chelates zinc and copper acts as an antioxidant. The main biochemical change induced in some studies by clioquinol was a marked reduction in lipid peroxidation at all time points (Bareggi SR et al. 2009).

RMH Note: I believe that clioquinol is another example of dead bodies from antioxidant use.

One thousand and thirty-one longstanding patients with subacute myelo-optico-neuropathy (SMON; 275 males, 756 females) were examined in 2002, **32 years after banning of clioquinol.** About 41% of patients were still **difficult to walk** independently, including 15.8% of completely loss of locomotion. One point six percent of patients were in **complete blindness** and 5.8% had severe visual impairment. As for complications, a **high incidence was revealed with cataract (56.2%), hypertension (40.2%),** vertebral disease (35.5%), and limb articular disease (31.5%).

These results indicate the serious sequelae of clioquinol intoxication, SMON (Konogaya et al. 2004).

RMH Note: I believe that this supports my view that an EMOD insufficiency is, in part, causative of cataract formation. This is the effect of the antioxidant clioquinol.

Tragic Deaths of 38 Infants by Lethal IV Antioxidant Vitamin E

Also, please remember the **fatal syndrome characterized by progressive clinical deterioration with unexplained thrombocytopenia, renal dysfunction, cholestasis, and ascites developed in certain infants throughout the United States who had received E-Ferol, an intravenous vitamin E supplement. (THE TRAGIC CASE HISTORY OF INTRAVENOUS VITAMIN E (The New York Times) May 27, 1984 By PHILIP M. BOFFEY).**

Prolonged exposure to E-Ferol was associated with progressive intralobular cholestasis, inflammation of hepatic venules, and extensive sinusoidal veno-occlusion by fibrosis. E-Ferol, contained 25 units per milliliter of dl-alpha-tocopheryl acetate solubilized with 9% polysorbate 80 and 1% polysorbate 20. They proposed that vasculocentric hepatotoxicity is the basis for the observed clinical syndrome that represents the cumulative effect of one or more of the constituents of E-Ferol (Bove et al, 1985).

All affected infants received E-Ferol; some affected infants received up to 1 ml or more daily. **Both outbreaks ceased shortly after use of E-Ferol was discontinued. Three were jailed for selling drug (Vitamin E) that killed 38 babies.**

The **Center for Drug Evaluation and Research, Food and Drug Administration, Rockville, Maryland, concluded that the use of E-Ferol in these neonatal intensive care units was associated with increased morbidity and mortality among exposed infants** (Arrowsmith et al, 1989).

Research has shown infants who received E-Ferol injections are at an increased lifetime risk for reproductive problems, cervical and vaginal cancer, and other health problems.

RMH Note: Here is another basic point: the increase in disease risk and mortality seen with the antioxidant vitamins cannot be expected to increase the life span. Obviously, they would be expected to decrease the life span.

RMH Note: Additionally, for those that say that there have been only a "few" negative studies with the antioxidant vitamins, there are now hundreds (over 500) which I have compiled and reported on. That is, indeed, shocking!

The Arrival of Antioxidants

Dr. Howes' Antioxidant Hype Update 2017

August 17, 2017. The Ponchatoula Times.

"The notion that antioxidant supplements are the potential cure for all of our maladies is pure nonsense, non-science."

By Dr. Randolph M. Howes, MD PhD.

Times Columnist

Antioxidant hype has been with us for decades, but scientific evidence has collectively demonstrated this hype to be disingenuous, or outright misleading.

For decades, I have extensively researched the subject of antioxidants. I discovered that in over 500 scientific studies, antioxidants have proven to be ineffective and frequently harmful (please check out my book on amazon.com entitled "Antioxidants Linked to Deadly Unintended Consequences").

Overall, there is no evidence from randomized controlled trials that any specific diets or dietary supplements prevent or treat cancer, heart disease, Alzheimer's disease, dementia or strokes. This should be expected since our very survival depends on prooxidant biochemistry, not antioxidant biochemistry.

Our bodies fight bacteria, fungi, protozoans, viruses, a wide variety of pathogens and cancer prooxidatively, and not with antioxidants. Nearly all of our energy comes from normal oxidative biochemistry.

Still, according to the National Institutes of Health, more than half of adults in the U.S. consume some kind of antioxidant product, spending $37 billion each year.

Via oxidation, our bodies keep us supplied with protective oxygen free radicals or so called reactive oxygen species (ROS). In fact, scientist Barry Halliwell says, "One per cent or more of the oxygen we consume turns into ROS. Over a year, a human body makes 1.7kg of ROS."

Yet, some uninformed or misinformed investigators still cry out the false mantra of antioxidants.

Evidence gathered over the past few years shows that, at best, antioxidant supplements do little or nothing to benefit our health. At worst, large doses have been shown to have the opposite effect, promoting the very problems they were supposed to stamp out.

The suggestion that antioxidant supplements can prevent chronic diseases has not been proved or consistently supported by the findings of published intervention trials.

Scientific data does not justify the use of antioxidant vitamin supplements for cardiovascular disease (CVD, heart disease) risk reduction.

This position is consistent with recommendations that have been made by the American Heart Association (AHA) in 2004 for the prevention of CVD in women as well as by the American College of Cardiology and AHA in 2002 for patients with chronic stable angina.

No consistent data suggest that consuming micronutrients (including antioxidant supplements) at levels exceeding those provided by a dietary pattern consistent with AHA Dietary Guidelines will confer additional benefit with regard to CVD risk reduction.

Large clinical trials with antioxidants failed to demonstrate any benefit for diabetic patients.

We must follow the scientific data to find the truth. We should stop wasting our hard-earned money on mythology and hype, all in the name of profit.

The notion that antioxidant supplements are the potential cure for all of our maladies is pure nonsense, non-science.

On the basis of the free radical theory, large-scale multicenter clinical trials of antioxidant supplementation were carried out. However, antioxidant supplementation not only failed to benefit patients but was associated with a significant increase in cancer incidence and possibly mortality.

Dietary supplementation with vitamin E, for example, significantly increases the risk of prostate cancer in healthy adults. Another study showed that dietary supplementation with vitamin E and beta-carotene (a vitamin A precursor) increases the incidence of lung cancer in male smokers.

So, what are we doing when we take megadoses of antioxidant supplements?

So now we can explain the seemingly paradoxical observation that although ROS can cause mutations, inhibiting them actually promotes tumor survival and metastatic spread. And as we know, it is not the primary tumor that kills that patient. It is the metastases to distant organs.

You probably wonder, "Yes, antioxidants promote metastases of pre-existing tumors." But what happens in healthy people?

The answer is twofold. First, cells undergo mutation all the time, day in and day out. Fortunately, thanks to the natural metabolic and immunological defenses perfected by eons of natural selection work, the vast majority of such cells are eliminated. But no system is perfect, and every so often some cells escape the defense mechanisms and proliferate. These nascent tumors will welcome an added boost to their proliferation provided by antioxidants.

Second, it has been shown that about 30% of males over 50 have resting prostate tumor cells when they die of unrelated causes. Is it any wonder then that the antioxidant vitamin E, promoted as a preventative of prostate cancer, actually promotes it?

Elena Piskounova and her coworkers studied human melanoma cells in mice to investigate the oxidative stress and the antioxidant system in these cells. She found that as a tumor starts to grow, the elevated metabolic activity of the mutated cells, as well as their uncontrolled proliferation, displaces cells from their normal environmental niches, and that can cause enhanced production of EMODs, reactive oxygen species (ROS).

These highly reactive molecules create a state of oxidative stress that can kill the cells. Successful tumor cells combat this stress by producing antioxidant molecules that neutralize EMODs.

However, tumor cells that detach from a primary tumor experience high EMOD levels again, both in the bloodstream and at distant sites where they attempt to form metastatic tumors. The increase in oxidative stress should kill those cells, and indeed most of them die. But the few successful tumor cells that establish metastases have undergone metabolic changes that allow them to better withstand oxidative stress (EMODs).

In a commentary on the melanoma study in Nature, the authors conclude: "These findings, taken together, necessitate a revised perspective of the role of EMODs in cancer, and force us to view it as an unlikely ally in our quest to treat this deadly disease."

**EMODs (electronically modified oxygen derivatives)
are not as harmful as you might've imagined,
not as destructive as you might've thought
and not the "inner enemy" you might've feared.
In fact, your life depends on them.**

R. M. Howes, M.D., Ph.D.

5-28-10

Antioxidant Supplements are Everywhere

Antioxidants primarily gained public attention in the 1990s.

It is estimated that about half the U.S. adult population takes dietary supplements. The increase in the number of nutritional supplements has assumed epidemic proportion from 4,000 in 1994 to 55,000 in 2012.

In 2007, out-of-pocket expenditures for herbal or complementary nutritional products reached $14.8 billion. Obviously, an unregulated lucrative business, inviting both fraud and enabling senators, the likes of Republican Senator Hatch (UT) and Democratic senator Harkin (IA), who authored legislation to prohibit meaningful oversight by the FDA.

Why would these public servants do something so patently harmful to the public? Follow the money, it always works!

As long as the problem was 'only' ~23,000 emergency department visits per year attributed to adverse events related to dietary supplements, it was 'no big deal'. But **now that the relationship to cancer deaths is not just an association but causal,** are we finally going to do something about it?

Antioxidant therapies have been evaluated in placebo-controlled trials involving tens of thousands of patients. Despite pathophysiologic, epidemiologic, and mechanistic data suggesting otherwise, these clinical trial results have been, to date, mostly negative in the setting of chronic preventative therapy.

Antioxidant therapies have been evaluated in placebo-controlled trials involving hundreds of thousands of patients. Despite pathophysiologic, epidemiologic, and mechanistic data suggesting otherwise, these clinical trial results have been, to date, mostly negative in the setting of chronic preventative therapy. (Steinhubl, 2008)

Even conservative and responsible scientists dared dream that antioxidants might cure everything from Alzheimer's disease to old age. Antioxidants may one day play an important role in the treatment of certain diseases or in routine health maintenance, but so far there is insufficient evidence to make any confident predictions or to make specific recommendations.

Before scientists could even write their research grants, however, there emerged from this "irrational exuberance" a cottage industry of antioxidant products marketed as if they were the very elixir of life. But two decades later a complex picture has emerged – and continues to emerge. At this point what does the evidence say about oxidative stress and antioxidants?

So, if anything, the basic science plausibility of antioxidants has increased over time. There is no question that oxidative damage is occurring as part of neurodegenerative diseases. But biological symptoms are frustratingly complex and often mock our simplistic explanations. It is possible, for example, that oxidative damage may be more of a consequence of the cellular damage that occurs than a cause of it. Therefore, antioxidants may not necessarily stop the cellular damage.

Medicine also seems to follow the law of unintended consequences. For example, it is also possible that antioxidants may decrease the beneficial actions of some oxidative agents.

Ultimately, we need clinical evidence to make any conclusions about the health benefits, if any, of nutritional antioxidants.

In general research into antioxidants as treatment for the various neurodegenerative diseases has been disappointingly negative.

The use of antioxidant supplements and the consuming of foods high in natural antioxidants for routine health maintenance needs to be considered completely separately from the use of antioxidants in treating specific diseases. But here the evidence is also ambiguous. The difficulty here is that most studies are epidemiological and don't control for many variables, such as the effects having a generally healthful diet. **A Danish meta-analysis published one year ago in JAMA even showed that the use of certain antioxidants was associated with a higher mortality rate**. All of this evidence, good and bad, is preliminary. Long term, prospective, definitive trials are simply lacking.

Antioxidants work by generously giving electrons to free radicals.

There are hundreds, probably thousands, of different substances that can act as antioxidants. The most familiar ones are vitamin C, vitamin E, beta-carotene, and other related carotenoids, along with the minerals selenium and manganese. They're joined by glutathione, coenzyme Q10, lipoic acid, flavonoids, phenols, polyphenols, phytoestrogens, and many more.

Even before the results of these trials were in, the media, and the supplement and food industries began to hype the benefits of "antioxidants." Frozen berries, green tea, and other foods labeled as being rich in antioxidants began popping up in stores. Supplement makers touted the disease-fighting properties of all sorts of antioxidants.

The trials were mixed, but most have not found the hoped-for benefits. Most research teams reported that vitamin E and other antioxidant supplements didn't protect against heart disease or cancer. One study even showed that taking beta-carotene may actually increase the chances of developing lung cancer in smokers.

The studies so far are inconclusive, but generally don't provide strong evidence that antioxidant supplements have a substantial impact on disease.

These mostly disappointing results haven't stopped food companies and supplement sellers from banking on antioxidants. Indeed, antioxidant supplements represent a $500 million-dollar industry that continues to grow. Antioxidants are still added to breakfast cereals, sports bars, energy drinks, and other processed foods, and they are promoted as additives that can prevent heart disease, cancer, cataracts, memory loss, and a host of other conditions.

Often the claims have stretched and distorted the data: While it's true that the package of antioxidants, minerals, fiber, and other substances found naturally in fruits, vegetables, and whole grains helps prevent a variety of chronic diseases, it is unlikely that high doses of antioxidants can accomplish the same feat.

Attempts to translate our understanding of these molecular and cellular mechanisms into clinical practice have failed so far. A vast number of clinical trials report no benefit of antioxidant supplementation in the genesis or progression of cardiovascular diseases (Steven R Steinhubl, 2008)

Taken together, the negative results clearly indicate that the use of antioxidant drugs in clinical applications requires precise knowledge of their physiological actions and underlying molecular mechanisms to understand their beneficial and avoid potential adverse effects.

False Claims Exposed

John Ioannidis has famously blown the whistle on how much of what is published in even the leading medical journals is simply wrong. (Lundberg, 2015) (Harris, 2017)

There are far too many journals with far too many articles lacking in scientific validity.

This can destroy the true value of evidence-based medicine.

Is it just the research or is it also the so called 'expert committees' that produce the guidelines that really enshrine bad research into law?

Can we also attribute the plethora of publications to the "publish or perish" mentality which prevails in our ivory tower institutions, that rewards publication regardless of the worth of the data or its conclusions?

Americans will swallow anything

Section Two: Cancer and XS Antioxidants

Cancer is the leading cause of death across the globe.

Antioxidants prove troublesome against cancer

Antioxidants May Make Cancer Worse - Scientific American
https://**www.scientificamerican.com**/article/**antioxidants**-may-make...

Antioxidants May Make Cancer Worse. ... of the **antioxidant** beta-carotene increased the risk of lung ... treated mice caused levels of RhoA to **increase** in their ...

Antioxidants may INCREASE cancer risk - Daily Mail Online

www.dailymail.co.uk/health/article-3265548/**Antioxidants-INCREASE**...

Compounds hailed for their **cancer**-fighting abilities could in fact **increase** the **risk** of the disease, experts have warned. **Antioxidants** could double the rate at which ...

Cancer myths about antioxidant supplements need to die ...

https://**arstechnica.com**/science/2015/10/myths-about-**antioxidant**...

And, this month, two studies add to evidence that antioxidants may actually increase the spread and severity of some cancers. Antioxidants and cancer. When cancer patients first get their diagnosis, many hit the internet to learn more, Martin Bergö, a molecular medicine ...

Antioxidant Supplements Don't Fight Cancer, Research Suggests

https://**www.livescience.com**/46927-**antioxidant-supplements-cancer**.html

Antioxidant Supplements Don't Fight **Cancer**, ... the possible benefits of **antioxidants** in lowering **cancer risk**, ... an 18 percent **increased** rate of lung **cancer**, ...

Antioxidant Supplements Fuel Lung Cancer in Mice

www.cancer.org › ... › Latest News

Antioxidant Supplements Fuel Lung Cancer in Mice. ... **Antioxidants**, lung **cancer**, ... "If you have lung **cancer** or an **increased risk** of developing lung **cancer**, ...

Antioxidants can increase melanoma metastasis in mice ...

stm.sciencemag.org/content/7/308/308re8

Antioxidants in the diet and supplements are widely used to protect against cancer, but clinical trials with antioxidants do not support this concept. Some trials show that antioxidants actually increase cancer risk and a study in mice showed that antioxidants accelerate the progression of primary lung tumors.

- Published in:
Science Translational Medicine · 2015
Authors:
Le Gal K · Mohamed X Ibrahim · C Wiel · Volkan I Sayin · Murali K Akula · Camilla Kar...
Affiliation:
University of Gothenburg

Antioxidant Supplements May Accelerate Melanoma Spread ...

https://**www.huffingtonpost.com**/entry/**antioxidant-cancer**-spread_us...

Oct 08, 2015 · In addition, other research has shown that taking antioxidant supplements may actually increase the risk of developing certain cancers in otherwise healthy people. For example, in one study of 36,000 healthy, middle-age men, researchers found that taking vitamin E for several years was linked to a 17 percent increase in the men's risk of prostate cancer.

Antioxidants speed up lung cancer growth, study shows ...

www.foxnews.com/health/2014/01/29/**antioxidants**-speed-up-lung...

Jan 29, 2014 · Testing the popular theory that **antioxidants** prevent **cancer**, ... **Cancer Antioxidants speed up lung cancer** ... **increased risk** of prostate **cancer** if ...

The latest study about antioxidants is terrifying ...

https://**www.washingtonpost.com**/news/to-your-health/wp/2015/10/14/...

A similar study conducted at Vanderbilt University and published in PLoS One in 2012 involving mice with prostate **cancer** also showed that **antioxidants** ... **increase** in ...

Antioxidants Might Accelerate Cancer - Business Insider

www.businessinsider.com/**antioxidants**-might-accelerate-**cancer**-2014-1

Basic Cancer Considerations

This information was taken from a Reuters article by Will Dunham, December 17, 2007.

About **7.6 million people will die this year worldwide from various types of cancer**, with lung cancer -- heavily driven by smoking -- killing 975,000 men and 376,000 women, the American Cancer Society said on Monday.

In all, about **12.3 million people will develop cancer this year**, the organization projected, using data from the International Agency for Research on Cancer, a branch of the World Health Organization.

About **20,000 people die of cancer every day worldwide**, the report showed. Smoking was heavily responsible for the lung cancer scourge. (**13.8/minute die from cancer worldwide or 833/hour).**

Cancer's burden is on the rise in developing countries as deaths from infectious diseases and child mortality fall and more people live longer, American Cancer Society epidemiologist Ahmedin Jemal said in a telephone interview. Cancer is more common as people get older, Jemal noted.

Cancer also is increasing in developing countries as people embrace **habits linked to cancer such as smoking and fattier diets,** Jemal said.

The report estimated 5.4 million people will get cancer and 2.9 million will die of cancer in developed nations, with 6.7 million cases and 4.7 million deaths in developing nations.

Overall, 75 percent of children with cancer live for five years in Europe and North America, compared to three-year survival rates of only 48 to 62 percent in Central American countries.

Cancers related to infections, such as stomach, liver and cervical cancer, were more common in developing countries, the group said. Fewer people survive cancer in developing countries due to lack of availability of early detection and treatment services, according to the report.

Globally, 15 percent of all cancers are caused by infections. The Helicobacter pylori bacteria causes stomach cancer, human papillomavirus causes cervical cancer and hepatitis can cause liver cancer.

Among men, the three most commonly diagnosed cancers are prostate, lung and colorectal cancer in developed countries and lung, stomach and liver cancer in developing countries.

Among women, the three most common cancers are lung, breast and colorectal in developed countries and breast, cervical and stomach cancer in developing countries.

About 465,000 women will die of breast cancer this year, making it the leading cause of cancer death among women worldwide.

THE FREE RADICAL FALLACY

Diseases Related to Oxygen-Derived Free Radicals: An Erroneous Concept

Oxygen-derived free radicals are very important mediators of cell injury and death. Not only are these highly reactive chemical species important in the aging process, but they are either directly or indirectly involved in a wide variety of clinical disorders, such as atherosclerosis, reperfusion injury, pulmonary toxicity, macular degeneration, cataractogenesis, and cancer. In addition, they play an important role in chronic granulomatous disease and act as secondary sources of cellular injury in chronic inflammatory processes and several disorders of the central nervous system. Furthermore, a wide variety of drugs and xenobiotics are themselves either converted to, or stimulate the formation of, free radicals. Prevention and/or treatment of many of these disorders may be possible by appropriate antioxidant therapy, either currently available or to be discovered through continued research. **This is entirely speculative and lacks any proof. Certainly, oxygen products can be linked to many metabolic events, but they have not been proven to be causative of any disease and antioxidants have not been proven to prevent or reverse any of these same diseases. All of it is speculation.** (Knight, 1995)

My EMOD Insufficiency Theory for the Allowance of Tumorous Growths

One of the primary motivating factors in my investigations into oxygen metabololology was an awareness of spontaneous healing or regression of cancer. The fact that any cancers would spontaneously regress told me that the human body had the capacity to cure cancer. Spontaneous regression of cancer had been known for over a century, but explanatory biochemical mechanisms were lacking. My intense studies of EMODs showed me that they held the key to cancer therapy and prevention.

This JAMA article illustrates my same line of reasoning (*JAMA*. 2007;298:344. JAMA 100 Years Ago Section Editor: Jennifer Reiling, Assistant Editor):

THE SPONTANEOUS HEALING OF CANCER. July 20, 1907 (*JAMA*. 1907;49:249-250)

At the meeting of the international conference for the investigation of cancer, in Heidelberg last year, Czerny discussed, from his rich experience, the question of the healing or retrogression of cancers which have seemed to be inoperable (Reported in the Ztschr. f. Krebsforschung, 1907, v, 27).

A hundred years ago the Society for the Investigation of Cancer, in London, propounded for consideration the question: Can cancer ever undergo a natural healing? At the present time (1907) **it can scarcely be denied that the answer must be in the affirmative, in view of carefully reported cases in which cancers have disappeared after inadequate treatment or none at all**; yet such cases are among the rarities of medical literature.

Surely, if cancers do ever cease to grow, the reason for this cessation of malignant propensities **should be the object of a most zealous search**. Czerny recalls several instances in the literature in which persons with cancers too far advanced for complete removal have, to the surprise of all, improved greatly after palliative local procedures, and, in a few cases, have shown no recurrence of the growth and no development of metastases. He quotes, with evident approval, the statement of Lomar, that **such a favorable outcome is much more likely to follow local treatment with the thermocautery or with chemical caustics** than incomplete operations with the knife. **I believe that this is due to the massive injury or damage caused by these agents and the consequent EMOD reaction of the body to heal and repair itself.**

Cases in which spontaneous healing has occurred without operative interference of any kind are much more rare, even, than these. Czerny recalls one case in which a patient with mammary carcinoma developed an **erysipelatous infection** in the wound of the second operation for recurrent growths, and who is now, twenty years later, free from all signs of the disease. He has also observed instances in which the microscope has shown that the line of incision for the removal of intestinal cancer has passed through cancerous tissue, and yet the patients have lived many years without recurrence of the growth. Among his cases of gastroenterostomy for inoperable cancer of the stomach have been many in which the growth decreased greatly in size after the operation, and an assistant was able to collect from the hospital records nine cases in which the patients have survived the operation over four years and are now in good health.

This too, I believe is due to the body's EMOD and PMN response and the generation of high levels of EMODs, which the body generates secondary to the trauma of surgery. However, surgery can "seed" the body with cancer but it must rest in "fertile soil" to grow. If the area of rest has high EMOD levels, the metastatic cells will not be able to survive and will undergo apoptosis.

It is remarkable how differently **operative interference** affects malignant tumors. Often, especially with the slow-growing scirrhus cancer of the breast, it **seems to fan a smoldering flame of malignancy into a fiery outburst of consuming growth**; yet **sometimes the operation seems to stimulate, not the cancer cells, but the protective mechanisms**, whatever they may be, so that the retrogressive changes predominate over the proliferative, and the residue of the growth and its metastases disappear or remain quiescent for many years. Instructive cases have been observed in which persons have died from some unrelated affection several years after operation for cancer, and autopsy has revealed the presence in the internal organs and tissues of cancer nodules showing marked fibrotic changes with few visible cancer cells in the scar tissue. . . Undoubtedly such occurrences are not uncommon, both in patients showing unexpected recurrence several years after an apparently successful operation, and also in patients who have escaped recurrence, and in whom, therefore, the operator believes there could have been no deep metastasis at the time of operation.

If we could explain the reason for this occasional loss of proliferative power in metastatic growths after removal of the primary cancer, we might acquire knowledge of great value to therapeutic progress. Lubarsch, in his address before the same convention, recounts certain facts and theories that have a bearing on this problem (Ztschr. f. Krebsforschung, 1907, v, 117).

He recalls that Cohnheim found that **pieces of fetal periosteum, injected into the veins, formed nodules of bone where they lodged in the pulmonary arteries; but, as these nodules soon underwent destruction, he concluded that the formation of metastases does not depend solely on peculiar properties of the cancer cells, but also on inhibition of the normal resorption processes.** In carcinoma, Lubarsch believes, embolic tumor cells can not attain a foothold for growth unless poisonous substances are present in the blood stream which prevent the normal resorptive processes from destroying the cancer cells, and these poisonous substances come, presumably, from the primary growth.

I believe that an EMOD insufficiency "allows" the survival, development and growth of the cancerous cells. Thus, specific sites may be more favorable to metastatic growths if they have low levels of EMOD generation and subsequent low apoptotic inducement. This can also result from a systemic EMOD insufficiency in cases of immunodeficiency (HIV/AIDS) or diseases such as chronic granulomatous disease.

Further, I believe that the growth of metastatic cells is similar to the situation in which microbes are present in a commensal manner and as the host conditions change, they become pathogens. As was argued and ultimately agreed upon by Lister and Bernard, it is the milieu which is most important.

If Lubarsch's hypothesis is correct, it can be readily understood how, in some cases, removal of the primary growth could leave the metastases that have already formed in a condition under which they could not successfully resist the resorptive power of the surrounding tissues.

Blumenthal (Ztschr. f. Krebsforschung, 1907, v, 186) has found evidence that the autolytic enzymes of cancer cells differ from those of normal cells in being able to attack any sort of cell in the body, and that in advanced cancer these cancer enzymes exist free in the blood; possibly these are the agencies that Lubarsch believes prepare the soil for the successful growth of embolic cancer cells.

On the other hand, we can imagine that the checking of metastatic growths could result from a process of self-immunization. During the operative manipulation of the primary growth considerable quantities of cancerous material may be forced into the circulation, and the results might either be (1) a widespread dissemination of the disease, which is frequently observed, or (2) in case the resistance of the body were sufficient these cells would be destroyed, and in the process cause the formation of antibodies sufficient to enable the individual to overcome the secondary growths which had already obtained a foothold.

If our current concepts are accurate, development of neoplasia is potentially constantly present, and our apoptotic abilities are responsible for the killing of these continually forming entities. I believe that EMODs serve as the basis for the apoptotic execution

system. Agents which block or nullify EMODs, such as antioxidants, or agents which result in an EMOD insufficiency will create conditions which "allow" tumorous growth and proliferation.

I fervently believe that the potential therapeutic and preventative benefits of my UTOPIA theory and my ROSI syndrome research is unlimited.

A May 5, 1905 JAMA article stated that "The clinical course of cancer, internal as well as external, is variable. Periods of apparent recession follow periods of progress. Every physician and surgeon has observed that patients with cancer sometimes continue to live in comparative comfort even for years after the unquestioned development of metastatic tumors in vital organs. It is part of the natural history of the scirrhous form of carcinoma, for instance, to advance with comparative slowness at the outskirts while in the central part cicatricial shrinking equivalent in many cases to a kind of spontaneous healing is taking place. There is, indeed, very good reason to believe that while the human being "in the majority of cases falls victim to the invasion of carcinoma once started, yet it is not altogether without the calling into operation of opposing forces that this sad result is achieved." (*JAMA*.1905;44:1534, 1538-1539).

I find it curious that we have made only minimal advancements against cancer or its prevention over the past century. Obviously, we need creative and innovative ideas, such as I have presented over the past 10 years.

Additionally, I agree with the following quote: "There are examples throughout medicine where theory, extrapolation of effects through surrogate markers, and weaker levels of evidence have led to harmful consequences for our patients. We should focus on proven preventive therapies, such as blood pressure control, before advocating potentially harmful therapies." (Clarifying the evidence: vitamin E, vitamin A, and folate. G. Michael Allan, and Wendy Payne. Can Fam Physician. 2005 November 10; 51(11): 1471–1472).

Immune System Can Halt Growth of a Cancerous Tumor Without Killing It

A multinational team of researchers has shown for the first time (Nov. 19, 2007) that the immune system can stop the growth of a cancerous tumor without actually killing it.

Scientists have been working for years to use the immune system to eradicate cancers, a technique known as **immunotherapy**. The new findings prove an alternate to this approach exists: When the cancer can't be killed with immune attacks, it may be possible to find ways to use the immune system to contain it. The results also may help explain why some tumors seem to suddenly stop growing and go into a lasting period of dormancy. **I believe that this frequently happens when the EMODs hold the cancers in abeyance.**

The study appears in the advance online publication of *Nature* on Nov. 19, 2007. (Koebel et al, 2007)

"Thanks to the animal model we have developed, scientists can now reproduce this condition of tumor dormancy in the laboratory and look directly at **cancer cells being held in check by the**

immune system," says co-author Robert Schreiber, Ph.D., Alumni Professor of Pathology and Immunology at Washington University School of Medicine in St. Louis. "That will allow us to see if we can model this state therapeutically."

The study's authors call the cancer-immune system **stalemate equilibrium**. During equilibrium, the immune system both decreases the cancer's drive to replicate and kills some of the cancerous cells, but not quickly enough to eliminate or shrink the tumor.

"We may one day be able to use immunotherapy to artificially induce equilibrium and convert cancer into a chronic but controllable disease," suggests co-author Mark J. Smyth, Ph.D., professor of the Cancer Immunology Program at the Peter McCallum Cancer Centre in Melbourne, Australia. "Proper immune function is now appreciated as another important factor in preventing the development of some cancers. Further research and clinical validation of this process may also turn established cancers into a chronic condition, similar to other serious diseases that are controlled long-term by taking a medicine."

Scientists first proposed that the immune system might be able to recognize cancer cells as potentially harmful more than a century ago. Under a theory that came to be called cancer **immunosurveillance**, researchers suggested that if this recognition took place, the immune system would attack tumors with the same weapons it uses to eliminate invading microorganisms. Current immunotherapy efforts use therapeutic agents to increase the chances that the immune system will recognize and attack tumors.

But cancer immunosurveillance has been controversial. The theory had begun to fall out of favor over the years, and in 2001, Schreiber, graduate students Vijay Shankaran and Gavin Dunn, and Lloyd Old, M.D., director of the New York branch of the Ludwig Institute for Cancer Research, proposed a major revision. They called their new model cancer **immunoediting**.

Like the older theory, cancer immunoediting suggests that conflict between cancers and the immune system naturally takes place but proposes that three very different outcomes can result. The immune system can eliminate cancer, destroying it; the immune system can establish equilibrium with cancer, checking its growth but not eradicating it; or the cancer can escape from the immune system, likely becoming more malignant in the process.

Until this latest study, evidence for the second outcome was lacking. Schreiber, Smyth and their colleagues posited equilibrium's existence mainly based on other doctors' clinical experiences. Examples included **cancers that inexplicably go into remission for years. In addition, there have been hints that in a few cases organ transplants have transferred undetected dormant tumors to the recipients.**

I believe that both of these instances are readily explained by my EMOD insufficiency theory, ROSI syndrome

To directly observe dormant tumors in mice, researchers injected them with small doses of a chemical carcinogen. Mice that developed outright tumors were set aside; the remaining mice had small, stable masses at the site of the injections. **When certain components of these**

animals' immune systems were disabled, the small growths became full-blown cancers, suggesting that the immune system had previously been holding the tumors in check.

"We don't think the immune system has evolved to handle cancers," Schreiber notes. "Cancer is typically a disease of the elderly, who have moved beyond their reproductive years, so there probably was no evolutionary pressure for the immune system to find a way to fight cancer."

I disagree with this approach.

Schreiber, Smyth and Old speculate that from the immune system's point-of-view, a cancerous cell may look like a cell infected by an invading microorganism. To overcome the safeguards that prevent the immune system from attacking the body's own tissues, the tumor has to have a high level of immunogenicity, or ability to provoke an immune reaction. Cancer cells can reduce their immunogenicity by changing the materials they present to the immune system to more closely resemble those presented by normal tissue. This enables the third outcome of the immunoediting theory: escape.

Equilibrium sometimes may be a more common outcome of tumor-immune encounters than elimination. According to the researchers' theory, some of us may harbor dormant tumors that either developed spontaneously or from exposure to carcinogens. **They propose that these quiescent tumors are unleashed only as we age or are exposed to environmental, infectious or physical stresses that cause a breakdown of the immune system.**

To follow up, researchers plan a molecular-level investigation of what happens in tumors and the immune system during equilibrium. They also want to test their results' applicability both in humans and in different types of cancers.

"For example, we need to look at which tissues are regularly edited by the immune system and at how closely the immune system watches over these tissues," Schreiber says. **"If you completely knock out the immune system in mice, you'll see tumors spring up in some tissues but not in others, and this suggests that there may be differing levels of immune system monitoring in different tissue types."**

I believe that this suggests that there are varying EMOD levels in various tissues.

"Over the past decade, remarkable advances have been made in our understanding of how the immune system reacts against cancer and influences the course of the disease and defining the equilibrium phase of cancer immunoediting represents the newest milestone in these advances," says Old. "The challenge now is to incorporate these findings into our thinking about human cancer and to develop immunotherapeutic strategies that complement current methods of cancer treatment." (Koebel et al, 2007)

My Basic Observation on the Lack of Toxicity of Oxygen Free Radicals

Harman and his sycophants believe that oxygen and its EMODs are inherently toxic and responsible for the myriad of human diseases and aging. However, neither oxygen free radicals nor EMODs can be highly toxic.

First, we continually breathe and transport ground state triplet oxygen from the air (21% or one in five molecules in the air is diradical oxygen) across the alveolar membrane and into the red blood cell, from the moment of conception until our last breath. Ground state triplet oxygen is not only a free radical, it is a diradical and is essential for sustaining the life of aerobic organisms, such as humans.

O_2 is present in the human body in quantities exceeding that of all other elements combined. Any compound present in such high concentrations and continuously being taken into and carried throughout the body can hardly be considered to be toxic. Clearly, is there is any oxygen toxicity, it must be extremely low, if at all.

During metabolism for the production of energy, EMODs are generated in steady state levels and are markedly increased with health-rendering exercise. Ergo, EMODs must be of extremely low toxicity, if at all toxic in the living/breathing organism.

Howes EMOD Cancer Therapy

New Drug May Make Tumors Self-Destruct

Doctors are hopeful about **a new drug to treat skin cancer by causing tumor cells to self-destruct by overloading them with oxygen**.

Unlike regular cells, which naturally cannot have their oxidant levels raised beyond a certain threshold, **cancer cells cannot balance the amount of free radicals inside them**.

With the new drug **STA-4783**, doctors may be able to overload the cancer cells with oxygen-containing chemicals to the point where the cells cannot cope and simply die off, according to research presented 9/26/07 at a meeting of the European Cancer Organization in Barcelona.

"We are taking advantage of the Achilles heel of cancer cells," said Dr. Anthony Williams, vice president of clinical research at **Synta Pharmaceuticals Corp., based in Lexington, Massachusetts,** which paid for the study. **STA-4783, which has no effect on normal cells,** is the first of several such drugs planned for study, though no other companies have yet to release results from their research.

It could also be used against other cancers, such as pancreatic or ovarian, as they have been shown to naturally contain higher levels of oxygen. Because cancerous tumor cells already have high oxygen levels, they are easier to overload. Doctors said the study focused on skin cancer, though, as melanoma tumors were particularly deadly.

Williams detailed how the drug doubled the amount of time that advanced melanoma patients survived without their cancer worsening. The study followed 81 patients with serious skin cancer: 28 received the **standard chemotherapy drug paclitaxel** and lived an average of 1.8

months before their cancer worsened, while 53 got **paclitaxel** plus the new drug and lived an average of 3.7 months before the disease worsened.

Patients on the experimental combination survived an average of one year after being diagnosed while those getting only the standard treatment survived an average of 7.8 months. "This could have a profound effect on patients," said Dr. Alex Eggermont, president-elect of the European Cancer Organization and a surgical oncology professor at Erasmus University in Rotterdam, Netherlands. Eggermont was not linked to the study.

Because **STA-4783 targets only cancer cells**, Williams said the drug does not come with too many side effects. Less than 5 percent of patients suffered serious side effects, which were similar to those seen in regular chemotherapy treatments, such as a temporary lowering of white blood cells, back pain and fatigue.

Though the study showed patients living longer with the cancer, some doctors said they hoped that, because the drug caused tumor cells to die off, it could potentially be used as a cure.

"Melanoma is so phenomenally complex that we desperately need new drugs to fight it," Eggermont said. There are very few drugs available for people with advanced skin cancer, which kills 70 percent of patients within one year. Williams said scientific trials offered patients the best hope for living longer. "There are a few options that relieve the disease, but we have nothing that cures it," said Dr. Jorgan Bergh, a professor of oncology at Sweden's Karolinska Institute.

"This new drug is potentially interesting, but we still need to understand more about how it works and how that may interact with chemotherapy," he said. Synta Pharmaceuticals will soon start a bigger study with more than 600 patients at 150 cancer centers worldwide.

Other scientists have tried to provoke immune system attacks on skin cancer tumors, but that research is preliminary, and scientists have yet to find a permanent way to keep the immune system from attacking healthy cells as well.

"We need to try different, novel approaches to see what might work," Eggermont said, adding that the two methods — oxygen-overloading and boosting immune defenses — could even be combined. "If you remove the inhibitions on the immune system and combine that with a drug, then that could open the door to new treatments," he said.

Doctors emphasized that melanoma tumors were different from, and deadlier than, other cancers. For instance, nearly half of all patients with a skin cancer tumor that is 4 millimeters (0.2 inches) thick will die. But for a breast cancer patient with a tumor of the same size, nearly 95 percent of patients will live.

"Patients with advanced melanoma really do not have a lot of options," Williams said. "We only get one chance to save patients' lives, and this may be a good starting point."

This work is based on my EMOD therapy methods, which I have described in detail.

EMODs Kill Cancer

Reactive oxygen species are known to be potentially dangerous but are also **needed for signal-transduction pathways.** Tumor cells have relatively low amounts of superoxide dismutase (SOD), which quenches superoxide anion O_2^-, and as a result of a higher level of aerobic metabolism, higher concentrations of O_2^-, compared to normal cells. But this may not be true of all tumor cells. **Some tumor cells have relatively higher amounts of vitamin E, a potent anti-oxidant, and a higher level of anaerobic metabolism, resulting in a balance that is tilted more towards higher anti-oxidant capacity.**

In both instances of higher aerobic and anaerobic metabolism **methods designed to augment free radical generation in tumor cells can cause their death.** It is suggested that free radicals and lipid peroxides suppress the expression of Bcl-2, activate caspases and shorten telomere, and thus inducing apoptosis of tumor cells.

Ionizing radiation, anthracyclines, bleomycin and cytokines produce free radicals and thus are useful as anti-cancer agents. But they also produce many side-effects.

2-methoxyoestradiol and polyunsaturated fatty acids (PUFAs) inhibit SODs and cause an increase of O_2^- in tumor cells leading to their death. In addition, PUFAs (especially gamma-linolenic acid), 2-methoxyoestradiol and thalidomide may possess anti-angiogenic activity. **This suggests that free radicals can suppress angiogenesis.** Limited clinical studies done with gamma-linolenic acid showed that it can regress human brain gliomas without any significant side-effects. Thus, PUFAs, thalidomide and 2-methoxyoestradiol or their derivatives may offer a new radical approach to the treatment of cancer. (Med Sci Monit., 2002)

I believe that this is in conformity with my UTOPIA theory and of the fact that high levels of EMODs are needed to protect us from neoplasia.

Carcinostatic Activity of Inorganic Selenium Compounds Due to EMODs

All inorganic selenium compounds that express carcinostatic activity against cancer cells in vivo do so by interaction with thiol compounds and generation of free radical species. (Spallholz, 2001)

Mechanism of carcinostatic activity of selenium compounds:
There are a number of hypotheses that have been postulated to account for the experimental data that selenium prevents cancer.

Five hypotheses seem to be possible to account for selenium's chemopreventative activity. The five hypotheses postulated are: (1) selenium's antioxidant role as a component of the glutathione peroxidase enzymes, (2) selenium's enhancement of immunity, (3) selenium's effect on the metabolism of carcinogens, (4) selenium's interactions that affect protein synthesis and the cycle of cell division, and (5) the formation of anti-cancer selenium metabolites.

Unbelievably, they completely missed the fact that EMODs have great tumoricidal activity. They are still looking at its antioxidant character, which blocks EMOD tumoricidal activity, to explain its carcinostatic abilities.

Selenium properties

Selenium is an essential dietary nutrient for most animals and humans, which is incorporated into twelve or more known proteins or enzymes as an amino acid, selenocysteine. Most, but not all selenium from selenium compounds can be efficiently incorporated into selenocysteine. **The major dietary forms of selenium are L-selenomethionine from cereal and animal protein and L-selenocysteine from animal protein.** Fruits and vegetables generally contain very low levels of selenium. At supranutritional levels of dietary selenium, most but not all selenium compounds can reduce the incidence of naturally occurring, viral or chemically induced cancer in both animals and humans. A common characteristic of all selenium compounds that express significant experimental carcinostatic activity in vitro or in vivo is their interaction with thiols and the generation of free radical species. **Selenodiglutathione is the most potent selenium compound against cancer cells** and readily arrests their growth as compared to selenite and any other selenium compound. **At higher levels of dietary intake many selenium compounds can become toxic.**

In dealing with selenium we must remember the words of Paracelsus (1393-1441), "The dosage makes either a poison or a remedy."

Selenium toxicity was first confirmed in 1933 to occur in livestock that consumed plants of the genus **Astragalus**, Xylorrhiza, Oonopsis and Stanleya. However, in 1957, it was discovered that selenium was an essential nutrient for laboratory rats to prevent dietary necrosis and over a period of time it was found to be an essential nutrient for many mammalian species. In 1973, selenium was found to be a component of glutathione peroxidase in the form of selenoamino acid, selenocysteine. Dietary selenium from the inorganic salts and organic selenium compounds are metabolized into selenocysteine. **In 1988, the observation was made that oxidation of glutathione by selenite produced superoxide,** opening a new area for selenium research. (Spallholz, 1994)

Selenium superoxide production

Glutathione oxidase activity of selenium compunds

Superoxide is produced (toxic)	Superoxide is not produced (non-toxic)
Selenite	Elemental selenium
Selenium dioxide	Selenate
Selenocystine	Selenoethionine
Selenocystamine	Selenomethionine
Diselenodiproprionic acid	Selenobetaine

(Spallholz,1993)

Milner et al. showed with the help of an experiment that most carcinostatic forms of selenium are inorganic salts: selenite and selenium dioxide. Selenate, selenocystine and selenomethionine are less effective as carcinostatic agents in preventing tumor growth.

The experiment is shown in the following table and it can be seen that **even a small dose of selenite and selenium dioxide prevent tumor growth.**
Superoxide producing selenium compounds prevent tumor growth

Tumor incidence in mice injected with selenium compounds

Treatment	Experiment 1	Experiment 2	Experiment 3
Control	10/10 (infected/total-mice)	5/5	5/5
Inorganic forms	**Dose, µg/g weight**		
2		**1**	**0.25**
Selenium dioxide	0/10	0/5	1/ 4
Sodium selenite	0/10	0/5	0/3
Sodium selenate	0/10	0/5	2/5
Organic forms			
Selenomethionine	0/10	5/5	5/5
selenocystine	0/10	0/5	2/5

I believe that this is the same tumoricidal pattern that I have seen over and over again with EMOD producing compounds. In short, EMODs kill cancer and antioxidants block this activity.

No dietary supplement, including selenium, has proven useful so far for the prevention of cardiovascular disease or cancer in the general U.S. population.
The balance of the potential benefits and harms of selenium supplementation depends on the dietary selenium intake in different countries. However, the U.S. public needs to know that most people in this country receive adequate selenium from their diet. By taking selenium supplements on top of an adequate dietary intake, people may increase their risk for diabetes. (Bleys et al, 2007)

Introduction to low oxygen toxicity

It appears to me that the major diseases of cancer, atherosclerosis, diabetes, arthritis, stroke, HIV/AIDS, Alzeheimer's disease, hypertension and cataract formation, that they all share a common trait of premature aging. These diseases are commonly linked or clustered and I have endeavored to find the underlying biochemical mechanism responsible for this association. I have concluded that these diseases share a defect in oxygen metabolism, whereby they cannot produce adequate levels of electronically modified oxygen derivatives (EMODs).

Ergo, I believe that increasing the body's collective oxidative capacity will help provide disease prevention, prooxidant pathogen protection, increase tumoricidal activity and aid in oxidative self-healing. I refer to this EMOD deficiency as the "Howes' EMOD insufficiency syndrome." I have found overwhelming data to support my position.

Oxygen is essential to respiration, so oxygen supplementation has found use in medicine (as oxygen therapy). People who climb mountains or fly in depressurized airplanes sometimes have supplemental oxygen supplies; the reason is that increasing the proportion of oxygen in the breathing gas at low pressure acts to increase the inspired oxygen partial pressure nearer to that found at sea-level. A notable application of oxygen as a very low-pressure breathing gas, **is in modern spacesuits, where use of nearly pure oxygen at a total pressure of about 1/3rd normal, results in normal blood partial pressures of oxygen.**

Oxygen has seventeen known isotopes with atomic masses ranging from 12.03 u to 28.06 u. Three are stable, ^{16}O, ^{17}O, and ^{18}O, of which ^{16}O is the most abundant (over 99.7%). The radioisotopes all have half-lives of less than three minutes. At sea-level pressures, mixtures containing less than 50% oxygen are essentially non-toxic.

Oxygen is the most common component of the Earth's crust (49% by mass), the second most common component of the Earth as a whole (28.2% by mass), and the second most common component of the Earth's atmosphere (20.947% by volume), second to nitrogen.

Immune systems of higher organisms have long made use of reactive forms of oxygen which they produce. **Not only do antibodies catalyze production of peroxide from oxygen, it is now known that immune cells produce peroxide, superoxide, and singlet oxygen in the course of an immune response.** Recently, singlet oxygen has been found to be a source of biologically-produced ozone: this reaction proceeds through an unusual compound dihydrogen trioxide, also known as trioxidane, (HOOOH) which is an antibody-catalyzed product of singlet oxygen and water. This compound in turn disproportionates to ozone and peroxide, providing two powerful antibacterials.

High Levels of Antibodies. Low Levels of Cancer?

Active immunisation can stimulate the body to produce highly efficient IgE antibodies that attack tumors. This breakthrough, achieved in an animal model, is based on the skilful combination of two established experimental methods. The results are now being published in Cancer Research and are part of a project funded by the Austrian Science Fund FWF. The antibodies produced during the project belong to a class that also plays a key role in the development of allergic reactions. Consequently, the results will be a key focal point at the 2007 1st International AllergoOncology Symposium, which is to be held in Vienna, Austria.

I believe that this is the result of the IgE antibodies ability to generate and act via EMODs.

People who suffer from allergies are well acquainted with immunoglobulin E (IgE). It is this class of antibodies that plays a key role in causing an allergy sufferer's immune system to overreact. Oncologists too are very familiar with IgE. **Numerous in-depth studies have shown that those with raised levels of IgE are much less likely to suffer from certain types of cancer.** Or in other words, allergy sufferers are at a lower risk of developing cancer.

Inverse relationship of allergies and cancer

"In actual fact, the IgE produced during an allergic reaction does not attack cancer tumours but instead attacks allergens, for example pollen", explains the study's leader, Prof. Erika Jensen-Jarolim, head of the Department of Pathophysiology at the Medical University of Vienna. "The fact that IgE nevertheless acts against tumors is more of a fortunate side-effect of the highly efficient characteristic of this antibody class. It was our aim to make this antibody class, which is typical for allergies, act directly against tumours. At the same time, we wanted to encourage the long-term production of IgE in the body by means of active immunisation."

Prof. Jensen-Jarolim's group recently succeeded in achieving the latter active immunisation against certain types of tumor in mice. However, due to the selected type of immunisation (injection below the abdominal wall) the antibodies that were produced belonged to the **IgG class**. This type of antibody produces a much more limited and shorter-term effect against tumours than IgE antibodies.

I do not believe that the class of antibody is crucial, since all antibodies appear to go through the EMOD generation phase, according to Barbior's work.

Via the stomach (food allergies)

Prof. Jensen-Jarolim used one of her group's earlier successes achieved as part of another FWF project on **food allergies** to ensure that immunisation resulted in the intended activation of IgE. These findings prove that food proteins are effective in inducing IgE-dependent immune reactions when they withstand the acidic environment of the stomach.

Prof. Jensen-Jarolim's team therefore fed mice a peptide very similar to a tumor peptide while reducing acidification in the stomach, thereby hindering digestion of the peptide. As a result, a type of allergic reaction was triggered against this tumor-like peptide the mice produced tumor-specific IgE antibodies. **The result is the world's first active IgE-stimulating tumor vaccination.**

From Prof. Jensen-Jarolim's point of view, the publication of this work in Cancer Research comes at the perfect time just as the 1st International AllergoOncology Symposium starts on 16th April, 2007 in Vienna. This symposium, which she initiated and organised, will be the first time that specialists from the U.K., France, Italy, Canada, Austria and the U.S. have been brought together to analyse and discuss the links between allergies and cancer. The topics under discussion range from the use of allergic reactions to treat cancer to the application of mimotopes for active immunisation against cancer tumors. By providing ongoing support for their work, the FWF has also helped establish this new medical field in Austria and on a global scale.

This work is strongly supportive of my Unified theory. Also, the introduction of food allergies could be used in my combinatorial approach to increasing the oxidative capacity of the body to fight diseases, including cancer.

Inhibit mitochondrial respiration, increase EMODs and kill cancer

Cancer cells are under intrinsic increased oxidative stress and vulnerable to free radical-induced apoptosis. Pelicano et al. report a strategy to hinder mitochondrial electron transport and increase superoxide $O_2^{\cdot-}$ radical generation in human leukemia cells as a novel mechanism to enhance apoptosis induced by anticancer agents. This strategy was first tested in a proof-of-principle study **using rotenone, a specific inhibitor of mitochondrial electron transport complex I. Partial inhibition of mitochondrial respiration enhances electron leakage from the transport chain, leading to an increase in $O_2^{\cdot-}$ generation and sensitization of the leukemia cells to anticancer agents whose action involve free radical generation.** Using leukemia cells with genetic alterations in mitochondrial DNA and biochemical approaches, they further demonstrated that **As_2O_3, a clinically active anti-leukemia agent, inhibits mitochondrial respiratory function, increases free radical generation**, and enhances the activity of another $O_2^{\cdot-}$.-generating agent against cultured leukemia cells and primary leukemia cells isolated from patients. Their study showed that interfering mitochondrial respiration is a novel mechanism by which **As_2O_3** increases generation of free radicals. This novel mechanism of action provides a biochemical basis for developing new drug combination strategies using **As_2O_3** to enhance the activity of anticancer agents by promoting generation of free radicals. (Pelicano et al, 2003)

I can simply do this with my singlet oxygen generating system. This study shows that partial inhibition of the electron transport chain increases EMOD production. There is obviously a point at which inhibition of the ETC will reduce EMOD production.

SOD as possible target for killing cancer cells

Superoxide dismutases (SOD) are essential enzymes that eliminate superoxide radical $O_2^{\cdot-}$ and thus protect cells from damage induced by free radicals. The active $O_2^{\cdot-}$ production and low SOD activity in cancer cells may render the malignant cells highly dependent on SOD for survival and sensitive to inhibition of SOD. **Inhibition of SOD causes accumulation of cellular $O_2^{\cdot-}$ and leads to free-radical-mediated damage to mitochondrial membranes, the release of cytochrome c from mitochondria and apoptosis of the cancer cells.** Huang et al. results indicate that targeting SOD may be a promising approach to the selective killing of cancer cells, and that mechanism-based combinations of SOD inhibitors with free-radical-producing agents may have clinical applications. (Huang et al, 2000)

A super way to kill cancer cells?

Estrogen derivatives have been found to inhibit superoxide dismutase and selectively kill human leukemia cells *in vitro*. But is SOD a realistic target for human cancer therapy? (Halliwell, 2000)

Superoxide and other 'reactive oxygen species' have been implicated as contributors to tissue injury in a wide range of human diseases, and generation of excess reactive species can lead to cell death by apoptosis or, if the oxidative stress is severe, by necrosis. (Halliwell and Gutteridge, 1999)

Increased formation of reactive oxygen species also accompanies apoptosis induced by most, if not all, other stimuli (Hampton et al, 1998), **and free radical scavengers often (but not always) delay such apoptosis** (Hampton et al, 1995)

Apoptosis is now widely recognized as being a distinct process of importance both in normal physiology and pathology. In the current paradigm for apoptotic cell death, the activity of a family of proteases, caspases, related to interleukin-1 beta-converting enzyme (ICE) orchestrates the multiple downstream events, such as cell shrinkage, membrane blebbing, glutathione (GSH) efflux, and chromatin degradation that constitute apoptosis. Recent studies suggest that mitochondria could be the principle sensor and that the release of mitochondrial factors, such as cytochrome c, is the critical event governing the fate of the cell.--**One of the most reproducible inducers of apoptosis is mild oxidative stress, although it is unclear how an oxidative stimulus can activate the caspase cascade**. Oxidative modification of proteins and lipids has also been observed in cells undergoing apoptosis in response to nonoxidative stimuli, suggesting that intracellular oxidation may be a general feature of the effector phase of apoptosis. The caspases themselves are cysteine-dependent enzymes and, as such, appear to be redox sensitive. Indeed, our recent work on hydrogen peroxide-mediated apoptosis suggests that prolonged or excessive oxidative stress can actually prevent caspase activation. A physiological example of this is the NADPH oxidase-derived oxidants generated by stimulated neutrophils that prevent caspase activation in these cells. (Hampton et al, 1998)

Thus, many attempts are underway to develop treatments for human apoptosis-related diseases (especially for neurodegenerative diseases) using free radical scavengers. (Halliwell and Gutteridge, 1998)

Alternatively, if selective inhibitors of superoxide dismutases (SODs), enzymes that convert superoxide free radical ($O_2^{\bullet-}$) into hydrogen peroxide (H_2O_2) and oxygen (O_2), could be developed, cells would be rendered sensitive to free radical killing. In the September 21, 2000 issue of *Nature*, Huang *et al*. report that selective inhibition of SOD kills human cancer cells but not normal cells, suggesting that free radical-producing agents may also have clinical applications. (Huang et al, 2000)

There are two major forms of SOD in human cells—a mitochondrial isoform (manganese-containing SOD, MnSOD) and a copper-zinc containing SOD, CuZnSOD. CuZnSOD is primarily localized in the cytosol in mammalian cells, although some may be present in the nucleus, mitochondrial intermembrane space, lysosomes and peroxisomes. **Transgenic mice lacking MnSOD do not survive for long, suffering severe lung damage and neurodegeneration.** (Huang et al, 1999)

Phenotypic defects in CuZnSOD-negative knockout mice are more subtle, but both enzymes appear essential for healthy aerobic life.

HIV infection raises lung cancer risk

Independent of cigarette smoking, **infection with HIV, the virus that causes AIDS, is associated with an elevated risk for developing lung cancer**, a study shows. Dr. Gregory D. Kirk from the Bloomberg School of Public Health, Johns Hopkins University, Baltimore, Maryland and colleagues evaluated lung cancer deaths among participants in injection drug users followed since 1998 as part of an AIDS study.

There were 27 lung cancer deaths among the 2,086 participants; 14 of the deaths occurred in HIV-infected subjects. After adjusting for potentially confounding factors like age, sex and smoking status, **HIV infection was associated with a 3.6-fold increased risk for lung cancer death compared to HIV-negative status**.

"As HIV-infected persons survive longer, we are continuing to see that non-AIDS outcomes are becoming the primary causes of morbidity and mortality," Kirk told Reuters Health on 7/6/07. "Our study suggests that the risks for these non-AIDS outcomes may be modulated by HIV infection."

"We hope to combine our data with other HIV and at-risk cohort studies to confirm the association between HIV and lung cancer," Kirk said. "Also, we are evaluating a series of smoking/tobacco related biomarkers in HIV infected and uninfected persons with similar smoking patterns to compare if the biological effect of smoking differs by HIV status, and if so, is this related to degree of immune suppression or to antiretroviral treatment," he added. (Kirk, 2007)

I believe that this is an example of EMOD insufficiency resulting in increased occurrence of cancer, as I have predicted.

My take on cancer

Query: **Is there any cell type or organ or tissue which does not form cancer cells? If all cells can become cancerous, I believe that this means that all cells have proliferation held in check by high or apoptotic levels of EMODs. Thus, all cells are potentially immortal if the apoptotic check (governor) is removed by reducing EMOD levels. Ergo, cancer cells do not have to learn to become immortalized because that potential or property is always there but is held in check by EMOD levels and apoptotic control.**

Does telomerase affect EMOD production?

I predict that if it does, it will decrease EMOD levels and allow for uncontrolled proliferation. Reportedly, increasing telomere length assures continued cellular proliferation.

My concept of cancer differs in that the normal cell is constantly held in neoplastic check by EMODs and an insufficiency of EMODs allows the cell to express its inherent cancerous tendencies and proliferate uncontrollably. This does not require activating substances or growth inhibitor substances but only requires a deficiency of EMODs. Also, for growth it does not require additional growth factors but only requires circumstances of reduced EMOD production. Metastatic sites are those sites which have insufficiencies of EMODs which would normally trigger apoptosis and does not require the presence of growth stimulators.

I do not believe that cancer cells have to "learn" to invade and metastasize because all they have to do is to find another body area with reduced EMOD production. Sloughed cells of the primary site will normally be carried throughout the body, but they must find "fertile soil" for continued growth and sites with high apoptotic inducing EMOD levels will kill the metastatic cells. I have extended the fertile soil concept to apply to bacterial vegetations seen with prosthetic heart valves or with scarred valves.

Using the car metaphor suggested by Douglas Hanahan and Robert Weinberg, **I envision the cell as being a car with a reveved up engine and EMODs are acting like the brakes or the governor**. To go forward only requires the release of the brake, with an EMOD deficiency.

It is said that cancer cells have to "learn" to avoid the process of apoptosis. However, I believe that the ability to divide is already there and is always normally suppressed by EMODs. In other words, avoiding apoptosis does not have to be learned because the desire to proliferate is already inherently there in their cell.

Ascorbic acid tumoricidal effects via H_2O_2

Some of the following material was abstracted, excerpted or modified from: **Intravenous Ascorbate as a Tumor Cytotoxic Chemotherapeutic Agent.** (Riordan, et al, 1995)

Ascorbic acid and its salts (AA) are preferentially toxic to tumor cells in vitro and in vivo. Given in high enough doses to maintain plasma concentrations above levels that have been shown to be toxic to tumor cells in vitro, AA has the potential to selectively kill tumor cells in a manner similar to other tumor cytotoxic chemotherapeutic agents. Because of the similarities between normal and malignant cells — both being born of the same host — a chemotherapeutic dose which is cytotoxic to cancer cells can also be toxic to normal cells. Because infectious complications are one of the major causes of death in cancer patients, more host-non-toxic compounds — particularly compounds without immune suppressive qualities — need to be investigated for their chemotherapeutic value.

There is a 10 — 100-fold greater content of catalase in normal cells than in tumor cells. This potentially creates a large gap between the toxic dose for normal cells and for tumor cells of agents which induce hydrogen peroxide generation. Ascorbic acid and its salts (AA) are preferentially toxic to tumor cells in vitro and in vivo. **This preferential cytotoxicity has been demonstrated to be related to intracellular hydrogen peroxide generation.** (Benade et al, 1969) (Noto et al, 1989) (Cohen and Krasnow, 1987)

AA thus belongs in a class of substances which, given at the correct dosage, can preferentially induce cytotoxicity of tumor cells with negligible toxic effects to the host.

In 1969, researchers at the NCI reported AA was highly toxic to Ehrlich ascites cells in vitro. The goal of the study was to exploit the 10-100-fold lower catalase activity in tumor cells compared to normal cells. **The proposed cytotoxic mechanism was generation of toxic hydrogen peroxide. The toxicity was greatly enhanced by concomitant administration of 3-amino-1,2,4-triazole (ATA), a catalase inhibitor**. Catalase and glucose added to the culture medium and a low oxygen tension reduced the toxic effects of AA and ATA. The addition of vitamin K_3 (menadione sodium bisulfite) to the medium overcame the protective effects of low oxygen tension and glucose. In 1977, Bram et al reported preferential AA toxicity for several malignant melanoma cell lines, including four human-derived lines. They found that **catalytic concentrations of Cu^{2+} greatly increased the preferential toxicity for melanoma cells**. Another French group also found that AA and Cu^{2+} were toxic to mouse melanoma cells in vitro. Noto et al reported that AA plus vitamin K_3 had growth inhibiting action against three human tumor cell lines at non-toxic levels. Helgestad et al recently reported that new malignant T-cell line, isolated from a boy with malignant lymphoma, was very sensitive to AA in culture at concentrations achievable in human plasma. In 1980, Park reported that **several leukemic cell cultures were sensitive to AA at concentrations attainable in vivo**, while normal hemopoietic cells were not suppressed. **Metabolites of AA have also shown antitumor activity in vitro.**

AA is toxic to several types of human tumor cells at concentrations which are non-toxic to normal cells. The AA begins to reduce cell proliferation in the tumor cell line at the lowest concentration, 1.76 mg/dl, and is completely cytotoxic to the cells at 7.04 mg/dl, while significant inhibition of the normal cells is demonstrated only at a dose of 28.18 mg/dl and 100% cell death is realized only at a dose of 56.36 mg/41 (8-fold higher dose than the tumor cells). In addition, the normal cells grew at an enhanced rate at the low dosages (1.76 and 3.52 mg/dl). They also show preferential toxicity of AA for tumor cells. >95% toxicity to human endometrial adenocarcinoma and pancreatic tumor cells (ATCC AN3-CA and MIA PaCa-2) occurred at 20 and 30 mg/dl, respectively. **No toxicity or inhibition was demonstrated in the normal, human skin fibroblasts (ATCC CCD 25SK) even at the highest concentration of 50 mg/dl.**

Tsao et al reported that AA in the drinking water of mice significantly inhibited the growth of human mammary tumor fragment xenografts implanted in immunocompetent mice. **AA was also effective as a tumor inhibitor in this model when given in the diet along with cupric sulfate** (Tsao C S, Dunham W B, Ping Y L. In vivo antineoplastic activity of ascorbic acid for human mammary tumor. In vivo 1988; 2: 147-150). As a dietary additive alone, AA was not effective. This finding supports the theory that **the inhibitory action of AA is due to hydrogen peroxide (and hydroxyl radical) production** because of Cu^{2+}'s known ability to catalyze the production of these substances in the presence of AA. (Halliwell, Guteridge, 1989)

Cheap, safe drug dichloroacetate (DCA), kills most cancers

DaveScot: *31 January 2007*

It sounds almost too good to be true: a cheap and simple drug that kills almost all cancers by switching off their "immortality". The drug, **dichloroacetate (DCA),** has already been used for years to treat rare metabolic disorders and so is known to be relatively safe. It also has no patent, meaning it could be manufactured for a fraction of the cost of newly developed drugs.

Evangelos Michelakis of the University of Alberta in Edmonton, Canada, and his colleagues tested DCA on human cells cultured outside the body and found that **it killed lung, breast and brain cancer cells, but not healthy cells.** Tumors in rats deliberately infected with human cancer also shrank drastically when they were fed DCA-laced water for several weeks.

DCA attacks a unique feature of cancer cells: the fact that they make their energy throughout the main body of the cell, rather than in distinct organelles called mitochondria. This process, called glycolysis, is inefficient and uses up vast amounts of sugar. Until now it had been assumed that cancer cells used glycolysis because their mitochondria were irreparably damaged. However, Michelakis's experiments prove this is not the case, because DCA reawakened the mitochondria in cancer cells. The cells then withered and died (Cancer Cell, DOI: 10.1016/j.ccr.2006.10.020).

- Dichloroacetate
  - Dichloroacetate is the most potent stimulus of pyruvate dehydrogenase, the rate-limiting enzyme for the aerobic oxidation of glucose, pyruvate, and lactate. Dichloroacetate may inhibit glycolysis and, thereby, lactate production. Dichloroacetate also exerts a positive inotropic effect that has been attributed to improvement in myocardial glucose use and high-energy phosphate production.
  - The data from animal studies and one placebo-controlled double-blind clinical trial have shown that dichloroacetate was superior to placebo in improving the acid-base status of the patients; however, the magnitude of change was small and did not alter hemodynamics or survival.

Dichloroacetate (DCA) produces superoxide anion

Dichloroacetate (DCA) is one of the toxic by products that are formed during the chlorine disinfection process of drinking water. In this study, the developmental toxicity of DCA has been determined in zebrafish (Danio rerio) embryos. Embryos were exposed to different concentrations (4, 8, 16, and 32 mM) of the compound at the 4 h postfertilization (hpf) stage of development and were observed for different developmental toxic effects at 8, 24, 32, 55, 80, and 144 hpf. Exposure of embryos to 8-32 mM of DCA resulted in significant increases in the heart rate and blood flow of the 55 and 80 hpf embryos that turned into significant decreases at the 144 hpf time point. At 144 hpf, malformations of mouth structure, notochord bending, yolk sac edema and behavioral effects including perturbed swimming and feeding behaviors were also observed. **DCA was also found to produce time- and concentration-dependent increases in embryonic levels of superoxide anion (O_2^{*-}) and nitric oxide (NO), at various stages of development.** The results of the study suggest that DCA-induced developmental toxic effects in zebrafish embryos are associated with production of reactive oxygen species in those embryos (Dichloroacetate-induced developmental toxicity and production of reactive oxygen species in zebrafish embryos. E. Hassoun et al. J Biochem Mol Toxicol. 2005;19(1):52-8).

Dichloroacetate (DCA) superoxide production blocked by antioxidant, ellagic acid

The ability of **ellagic acid (EA)** to modulate dichloroacetic acid (DCA)-induced developmental toxicity and oxidative damage was examined in zebrafish embryos. Embryos were exposed to 20 mM EA administered concomitantly with 32 mM DCA at 4 hours postfertilization (hpf) and 20 h later. Embryos were observed through 144 hpf for developmental malformations, and production of superoxide anion (SA) and nitric oxide (NO) was determined in embryonic homogenates. **DCA was shown to produce developmental abnormalities and significant levels of SA and NO in zebrafish embryos.** EA exposure alleviated the developmental malformations observed in treated embryos and **decreased the levels of SA** and NO in those same embryos. Less than 10% of DCA + EA exposed embryos showed developmental malformations compared to 100% of embryos treated with DCA alone. Animals in this group that developed malformations were shown to have fewer defects than those treated with DCA only. Taken together, **the results confirm the involvement of oxidative stress in the developmental toxicity of DCA in zebrafish embryos and suggest possible protection against those effects with the use of antioxidants.** (Williams et al, 2006)

In short, I believe that reliable scientific data concerning potential dangers of antioxidant vitamins makes them tough pills to swallow.

Free radical theory of aging unsupported by human study

The hypothesis that the aging process is associated with mitochondrial dysfunction and oxidative stress has been investigated in human skeletal muscle. Muscle biopsy samples were taken from seven old male subjects [OS; 75 (range 61-86) years] and eight young male subjects [YS; 25 (22-31) years]. Oxidative function was measured both in permeabilised muscle fibres and isolated mitochondria. Despite matching the degree of physical activity, OS had a lower training status than YS as judged from pulmonary maximal O_2 consumption (Vdot; O_2 max, -36%) and handgrip strength (-20%). Both maximal respiration and creatine-stimulated respiration were reduced in muscle fibres from OS (-32 and -34%, respectively). In contrast, respiration in isolated mitochondria was similar in OS and YS. The discrepancy might be explained by a biased harvest of "healthy" mitochondria and/or disruption of structural components during the process of isolation. Cytochrome C oxidase was reduced (-40%, P<0.01), whereas UCP3 protein tended to be elevated in OS (P=0.09). **Generation of reactive oxygen species by isolated mitochondria and measures of antioxidative defence (muscle content of glutathione, glutathione redox status, antioxidative enzymes activity) were not significantly different between OS and YS.** It is concluded that aging is associated with mitochondrial dysfunction, which appears to be unrelated to reduced physical activity. **The hypothesis of increased oxidative stress in aged muscle could not be confirmed in this study.**
(Tonkonogi et al, 2003)

High EMODs and low toxicity in birds

Austad and University of Idaho ecologist Donna J. Holmes are looking skywar to explain aging. Five years ago they proposed birds as the ideal animal to use in aging studies. After all, birds are closer to humans, biologically speaking, than are worms or fruit flies, the favorite subjects of aging-study labs. They are **warm-blooded**, like us, so they don't lapse into periods of dormancy or hibernation, as do fish and turtles. Moreover, some birds live for decades against all odds (Why We Age. Steven N. Austad. John Wiley & Sons, 1997).

This is even more remarkable because, to rev up for flight, **birds generate extremely high levels of blood sugar**. The 150 parakeets twittering around a basement lab at the University of Idaho have blood sugar levels so high they should be diabetic. They have **elevated temperatures and burn energy at feverish rates**. Yet they live to 20, old for parakeets. These bird traits defy a primary theory of aging—**that increased metabolism creates higher levels of oxygen molecules, called free radicals, that oxidize cells, damaging tissue in ways normally associated with aging. I believe that this simple observation also invalidates the free radical theory of aging.**

Rather than rapidly growing weak and dying, birds carry on in good health, year after year. In 1998 Holmes, Austad and their colleagues reported that the cells of three bird species—canaries, European starlings and budgerigars (a.k.a. parakeets)—can endure a battery of oxidative stresses with surprisingly little damage. The scientists exposed these bird cells, along with the cells of mice, to **baths of hydrogen peroxide**, bolts of radiation, chambers of oxygen and doses of pesticide. Under these assaults, the DNA inside the mouse cells often unraveled, broke or stopped replicating, typical signs of free radical damage. The bird cells, on the other hand, **divided normally and repaired much of the induced DNA damage right away**. "We don't have any idea yet how the bird cells are doing it," Holmes says. "But it appears that birds have special enzymes that dispose of free radicals. If free radicals are a primary mechanism of aging, then this may explain why these birds live so long." **I believe that these are extremely important experiments. Bird cells are bombarded with strong oxidative species and they are proven to be of low toxicity, just as I have predicted.**

"Researchers used to believe that the older you get, the sicker you get," says Harvard Medical School physician Thomas T. Perls. "That's completely wrong." Perls has a few insights, gathered as head of the New England Centenarian Study, which tracks more than 450,000 older adults in Massachusetts to see who reaches 100 and why (THE QUEST TO BEAT AGING, Scientific American, 2000).

So far 169 centenarians have participated in the study; there is data on 250 others. They are a motley crew: **Some exercise. Some smoke. Some brazenly defy the notion of a healthy lifestyle. Nevertheless, almost all have lived free of cancer, and up to a fourth have escaped any form of dementia.**

I believe that this observation discounts the so called harmful effects of EMODs because these patients have generated countless quantities of EMODs and have remained cancer and Alzheimer's free for over a century. This occurred because EMODs are in fact of

low toxicity in the living/breathing cell. Otherwise, it could not have happened. According to the free radical theory, they should all be dead due to the stochastic accumulation of oxidative products over these many years of respiration and generation of EMODs. That did not happen.

Nature supplies ample evidence that the rate of aging is flexible rather than predetermined. The evidence comes from comparisons between species. A fruit fly lives three weeks, a mouse three years, a quahog clam 200 years, and a bristlecone pine 4,000 years. In each of these species, **the same cellular processes are at work**.

The rate-of-living model gives rise to some seductively simple ideas. It suggests, for example, that all species of mammal have the same number of heartbeats in a lifetime. And it was buttressed by evidence that the normal metabolic consumption of energy generates reactive molecules called **free radicals** that damage DNA, enzymes, and cell membranes. The damage accumulates over time and results in an organism's increased susceptibility to cancer, or its inability to repair clogged arteries, or a slide into senility. The free-radical model is now a leading theory of aging, and it fits neatly with the rate-of-living theory of life span: The faster the metabolism, the faster free radicals do their damage. **Yet, I have found exceptions to this line of reasoning repeatedly. Remember, exercise is good for you and it decreases the co-morbidity of cancer, atherosclerosis, diabetes, etc. (the EMOD insufficiency syndrome diseases).**

But the rate-of-living theory succumbed to the weight of exceptions. Birds, for example, have metabolisms twice as fast as those of mammals, yet they can live much longer. Parrots can outlive elephants; hummingbirds have been known to survive to 14—the equivalent, in terms of energy consumption per pound, of a human living to 500. A species of North American bat half as big as a mouse can live 30 years in the wild. Opossums, on the other hand, rarely last more than two years, even in captivity. There is one more glaring exception: **Humans live four times longer than they should based on their size and metabolic rate.** (DISCOVER Vol. 24 No. 11, November 2003).

The elephant with 30 heartbeats per minute lives for about 80yrs, while Harriet the tortoise with about 17 heartbeats per minute is still alive at 170yrs.

If one knowingly takes an antioxidant supplement which increases the incidence of cancer, they are indirectly committing suicide.

The following was taken from my book, Dangers of Antioxidants In Cancer Patients:

I have presented approximately 106 studies showing EMODs are responsible for induction of apoptosis.

I have presented approximately 50 studies showing that antioxidants can block EMOD induced apoptosis.

There are 26 studies === showing that antioxidants increased the risk of cancer and 2 showing they increased the mortality in cancer patients.

<u>**Antioxidants increased the risk of the following cancers:**</u>

- **lung cancer, (7 studies)** (up to 28% in smokers)
- **prostate cancer, (5 studies)**
- **doubled the risk of adenoma recurrence,**
- **squamous cell skin cancer (2 studies)**
- **melanoma**
- **higher rate of second primary cancers (head and neck), local   recurrence of the head and neck tumor tended to be higher, (2   studies)**
- **esophageal cancer, (2 studies)**
- **oral pre-malignant lesions,**
- **stomach cancer**
- **urinary bladder cancer**
- **breast cancer**

<u>**SUMMARY**</u> **of antioxidants causing various cancers include the following: lung cancer, breast cancer, prostate cancer, stomach cancer, esophageal cancer, head and neck cancer, bladder cancer, melanoma, adenomas, oral premalignant lesions.**

Antioxidants increased the mortality in the following cancer patient groups (2 studies):

- 1) increase overall mortality in gastrointestinal cancer patients,

- 2) esophageal cancer deaths increased 14% in patients, (2 studies)

26 STUDIES SHOWING RELATIONSHIP OF HARMFUL EFFECTS OF ANTIOXIDANTS TO CANCER:

1) Alpha-Tocopherol and beta-carotene supplements and lung cancer incidence in the alpha-tocopherol, beta-carotene cancer

prevention study: effects of base-line characteristics and study compliance. (Albanes et al, 1996) (#29,133 men, smokers) === Supplementation with alpha-tocopherol or beta-carotene does not prevent lung cancer in older men who smoke. *beta-Carotene supplementation at pharmacologic levels may modestly increase lung cancer incidence in cigarette smokers.* *The incidence of lung cancer was 18% higher among men who took the beta-carotene supplement and **eight percent more men in this group died, as compared to those receiving other treatments or placebo**.* (Albanes D., Heinonen O.P., Taylor P.R, 1996). Smokers should avoid high-dose beta-carotene supplementation.

2) **The β-Carotene and Retinol Efficacy Trial (CARET)** === (Omenn et al, 1996) (#14,254 heavy smokers and 4,060 asbestos workers) (total #18,314 men and women); randomized, double-blind, placebo-controlled intervention; Duration of Treatment Years: 4; Daily Dose: 30 mg β-carotene, 25,000 IU retinol (as retinyl palmitate); *28% increase in lung cancer; 26% increase in CVD (nonsignificant); 17% increase in total mortality* among treatment group. This study was stopped 21 months earlier than planned.

3) **Energy, nutrient intake and prostate cancer risk: a population-based case-control study in Sweden.** === (Andersson et al. 1996) (#1,062). *In age-adjusted analyses, there were positive associations of prostate cancer (all stages combined) risk with total energy intake as well as intake of total fat (saturated and monounsaturated), protein, retinol and zinc.* The positive association with energy intake was stronger for advanced cancer, with an excess risk of 70% for the highest quartile vs. the lowest. *After adjustment for energy intake, there was no apparent association of prostate cancers (all stages combined) with any of the investigated nutrients. However, a weak positive association between intake of retinol and advanced cancer was observed.* We conclude that our results provide some evidence that **total energy intake is a risk factor for prostate cancer**.

4) **Randomized Trial of Supplemental ß-Carotene to Prevent Second Head and Neck Cancer** === (Mayne et al, 2001) (#264 patients who had been curatively treated for a recent early-stage squamous cell carcinoma of the oral cavity, pharynx, or larynx.); randomized, placebo-controlled, double-blinded clinical trial; 50 mg of ß-carotene per day; **Supplemental ß-carotene had no significant effect on second head and neck cancer or lung cancer.** Whereas none of the effects were statistically significant, the **point estimates suggested a possible decrease in second head and neck cancer risk** *but a possible increase in lung cancer risk*.

5) **Selenium and vitamin E supplements for prostate cancer: evidence or embellishment?** === (Moyad et al. 2002) (# not available). **Selenium supplements provided a benefit only for those individuals who had lower levels of baseline**

plasma selenium. *Other subjects, with normal or higher levels, did not benefit and may have an increased risk for prostate cancer. Vitamin E supplements in higher doses (> or =100 IU) were also associated with a higher risk of aggressive or fatal prostate cancer in nonsmokers from a past prospective study.*

6) **Vitamins E & A fail to reduce incidence or mortality of lung cancer: Cochrane Database Syst Rev. 2003.** === (Caraballoso et al., 2003) (#109,394 participants); **When beta-carotene was combined with retinol, data from a single study showed that there was a statistically significant,** *increased risk of lung cancer incidence and mortality* **in people with risk factors for lung cancer who took both vitamins.**

7) **Neoplastic and Antineoplastic Effects of Beta Carotene on Colorectal Adenoma Recurrence: Results of a Randomized Trial** === (Baron et al, 2003) (#864 subjects who had had an adenoma removed and were polyp-free); *For participants who smoked cigarettes and also drank more than one alcoholic drink per day, beta carotene doubled the risk of adenoma recurrence.*

8) **Selenium supplementation and secondary prevention of nonmelanoma skin cancer in a randomized trial.** (Duffield-Lillico, 2003) (#1,312). **Results from the Nutritional Prevention of Cancer Trial conducted among individuals at high risk of nonmelanoma skin cancer continue to demonstrate that selenium supplementation is ineffective at preventing basal cell carcinoma and that** *it increases the risk of squamous cell carcinoma and total nonmelanoma skin cancer. Selenium supplementation was associated with statistically significantly elevated risk of squamous cell carcinoma.* (Duffield-Lillico, 2003).

9) **Use of multivitamins and prostate cancer mortality in a large cohort of US men.** === (Stevens et al, 2005) (#475,726 men who were cancer-free). **The death rate from prostate cancer was marginally higher among men who took multivitamins regularly** (> or =15 times/month) **compared to non-users; this risk was statistically significant only for those multivitamin users who used no additional (vitamin A, C, or E) supplements.** In addition, risk was greatest during the initial four years of follow-up. CONCLUSIONS: *Regular multivitamin use was associated with a small increase in prostate cancer death rates.*

10) **A randomized trial of antioxidant vitamins to prevent second primary cancers in head and neck cancer patients.** === (Bairati et al, 2005 Apr 6) (#540 patients with stage I or II head and neck cancer treated by radiation therapy).

Patients receiving alpha-tocopherol supplements had a higher rate of second primary cancers during the supplementation period but a lower rate after supplementation was discontinued. Similarly, *the rate of having a recurrence or second primary cancer was higher during but lower after supplementation with alpha-tocopherol.* The proportion of participants free of second primary cancer overall after 8 years of follow-up was similar in both arms. CONCLUSIONS: *alpha-Tocopherol supplementation produced unexpected adverse effects on the occurrence of second primary cancers and on cancer-free survival.* (Bairati et al, 2005 Apr 6).

Note: *Patients taking an antioxidant were 1.65 times more likely to suffer a return of their original cancer during the three years they were on the supplement. The risk was highest among those taking only vitamin E (1.86 times higher).* Five years after they stopped taking the supplement, their recurrence risk had fallen to the same level as those in the placebo group. **Although suggestive of harm, these results were not statistically significant.** ===

11) Randomized trial of antioxidant vitamins to prevent acute adverse effects of radiation therapy in head and neck cancer patients === (Bairati et al, 2005 Aug 20) (#540 patients with stage I or II head and neck cancer treated by radiation therapy) A randomized trial. During the course of the trial, **supplementation with beta-carotene was discontinued because of ethical concerns. Quality of life was not improved by the supplementation.** *The rate of local recurrence of the head and neck tumor tended to be higher in the supplement arm of the trial.* CONCLUSION: Supplementation with high doses of alpha-tocopherol and beta-carotene during radiation therapy could reduce the severity of treatment adverse effects. However, **this trial suggests that use of high doses of antioxidants as adjuvant therapy might compromise radiation treatment efficacy.**

Note: *Researchers were concerned to find that the rate of local recurrence (that is, a return of the original cancer) was 54 percent higher among patients on the combination pill than those on placebo.* There was a smaller but still worrisome increase among those on vitamin E only.

NCI COMMENT: "This is a large, well-done study with good compliance from the participants," said Eva Szabo, M.D., of the National Cancer Institute's Division of Cancer Prevention. "The results demonstrate that the use of vitamin E supplementation is not beneficial to patients with stage I or II head and neck cancer, either as a chemoprevention agent or to enhance quality of life during radiation therapy."

12) Smoking, alcohol drinking, green tea consumption and the risk of esophageal cancer in Japanese men. (Ishikawa et al, 2006) (#9,008 men in Cohort 1 and 17,715 men in Cohort 2) === *Cigarette smoking, alcohol drinking and green tea*

consumption were significantly associated with an increased risk of esophageal cancer. The population attributable fractions of esophageal cancer incidence that was attributable to smoking, alcohol drinking and green tea consumption were 72.0%, 48.6%, and 22.1%, respectively.

CONCLUSIONS: Among the variables studied, *smoking has the largest public health impact on esophageal cancer incidence in Japanese men, followed by alcohol drinking and green tea drinking* (Ishikawa et al, 2006).

13) **Multivitamin Use and Risk of Prostate Cancer in the National Institutes of Health–AARP Diet and Health Study** === (Lawson et al, 2007) (#295,344 men); investigated the association between multivitamin use and prostate cancer risk; *use of multivitamins more than seven times per week, when compared with never use, was associated with a doubling in the risk of fatal prostate cancer* (The study of Lawson et al. is observational, and therefore confounding by indication and other confounding cannot be excluded. But the sample studied is very large, which reduces random errors, and the study seems well conducted.)

14) **Health Professionals Follow-up Study (2007): Effect of vitamins C, E, A and carotenoids and the occurrence of oral pre-malignant lesions.** === (Maserejian et al, 2007) (#42,340 men enrolled in the Health Professionals Follow-up Study) (#207 found with oral premalignant lesions); researchers found no clear relationship with beta-carotene, lycopene, or lutein/zeaxanthin. **A trend for *increased risk of oral pre-malignant lesions was observed with vitamin E, especially among current smokers and with vitamin E supplements. Beta-carotene also increased the risk among current smokers.*** However, **dietary vitamin C was significantly associated with a reduced risk of oral premalignant lesions**: those with the highest intake had a 50 percent reduction in risk compared to those with the lowest intake.

15) **Antioxidant Supplementation Increases the Risk of Skin Cancers in Women but Not in Men.** === (Hercberg et al, 2007) (#French adults, 7,876 women and 5,141 men. Total # = 13,017). ***In women, the incidence of SC was higher in the antioxidant group.*** **Conversely, in men, incidence did not differ between the 2 treatment groups.** Despite the small number of events, *the incidence of melanoma was also higher in the antioxidant group for women.*

16) **National Institutes of Health State-of-the-Science Conference Statement: Multivitamin/Mineral Supplements and Chronic Disease Prevention (NIH State-of-the Science Panel. 2007).** === There is evidence, however, that certain ingredients in MVM supplements can produce adverse effects, including **reports from RCTs that noted excess lung cancer occurring in asbestos workers and smokers consuming β-**

carotene. In addition, *esophageal cancer excess was found with long-term follow-up of older Chinese patients (the Linxian study by Blot et al.) treated with selenium, β-carotene, and vitamin E supplements* (Blot et al, 1993) (NIH State-of-the Science Panel. 2007).

17) **Systematic review: primary and secondary prevention of gastrointestinal cancers with antioxidant supplements.** (Bjelakovic, Nikolova, Simonette and Gludd, 2008 Sept) === (#211,818 participants). We identified 20 randomized trials (211,818 participants) assessing beta-carotene, vitamin A, vitamin C, vitamin E, and selenium. **The antioxidant supplements were without a significant effect on the occurrence of gastrointestinal cancers. Antioxidant supplements had no significant effect on mortality in a random-effects model meta-analysis** but *significantly increased mortality in a fixed-effect model meta-analysis.* CONCLUSIONS: **There was no evidence that the studied antioxidant supplements prevented gastrointestinal cancers. On the contrary,** *they seem to increase overall mortality.*

18) **Long-term use of supplemental multivitamins, vitamin C, vitamin E, and folate does not reduce the risk of lung cancer. VITAL (VITamins And Lifestyle) study (2008)** (Slatore et al, 2008) === (#77,721 men and women); **There was no inverse association with any supplement.** *Supplemental vitamin E was associated with a small increased risk of lung cancer.* This risk of supplemental vitamin E was largely confined to current smokers and was greatest for non–small cell lung cancer.

19) **Efficacy of Antioxidant Supplementation in Reducing Primary Cancer Incidence and Mortality: Systematic Review and Meta-analysis.** === (Bardia et al, 2008) (#104,196 participants). *Beta carotene supplementation was associated with an increase in the incidence of cancer among smokers and with a trend toward increased cancer mortality.*

20) **Total and Cancer Mortality After Supplementation With Vitamins and Minerals: 10 year Follow-up of the Linxian General Population Nutrition Intervention Trial.** === (Qiao et al, 2009) (#29,584 adult participants) The General Population Nutrition Intervention Trial was a randomized primary esophageal and gastric cancer prevention trial. Treatment with **"factor D," a combination of 50 µg selenium, 30 mg vitamin E, and 15 mg beta-carotene showed** *esophageal cancer deaths increased 14% among those aged 55 years or older. Vitamin A and zinc supplementation was associated with increased total and stroke mortality.*

21) **Long-term use of beta-carotene, retinol, lycopene, and lutein supplements and lung cancer risk: results from the VITamins And**

Lifestyle (VITAL) study. === (Satia et al, 2009) (#77,126 (VITAL) cohort Study in Washington State). *Longer duration of use of individual beta-carotene, retinol, and lutein supplements (but not total 10-year average dose) was associated with statistically significantly elevated risk of total lung cancer and histologic cell types.*

22) **Folic acid and risk of prostate cancer: results from a randomized clinical trial.** (Figueiredo et al, 2009) (#643 randomly assigned men). === Among the **643 men who were randomly assigned** to placebo or **supplementation with folic acid,** *the estimated probability of being diagnosed with prostate cancer over a 10-year period was 9.7% in the folic acid group and 3.3% in the placebo group.*

23) **Green tea consumption and risk of stomach cancer: a meta-analysis of epidemiologic studies** (Myung, Int J Cancer. et al, 2009) (#13 epidemiologic studies) === *In the meta-analyses of the recent cohort studies, the highest green tea consumption was shown to significantly increase stomach cancer risk using the crude data,* **but no significant association between them was seen when using the adjusted data.** (Myung, Int J Cancer. et al, 2009).

24) **Green tea (Camellia sinensis) for the prevention of cancer. Chochrane Database Syst Rev. 2009 Jul 8;(3):CD005004).** (Boehm et al, 2009) (#Fifty-one studies with more than 1.6 million participants were included) === **There was limited to moderate evidence that the consumption of green tea reduced the risk of lung cancer, especially in men, and** *urinary bladder cancer or that it could even increase the risk of the latter.* **There is insufficient and conflicting evidence to give any firm recommendations regarding green tea consumption for cancer prevention.**

25) **Multivitamin use and breast cancer incidence in a prospective cohort of Swedish women.** === (Larsson et al, 2010) (#35,329 cancer-free women). *Multivitamin use was associated with a statistically significant increased risk of breast cancer. Use of multivitamins was linked to a statistically significant 19 per cent increased risk of breast cancer.*

Other research has found that *women who take multivitamins have increased breast density, which is linked to a relatively higher risk of breast cancer* (Berube et al, 2008). === **The widespread use of multiviamins points out an important public health concern.**

26) **Daily intake of antioxidants in relation to survival among adult patients diagnosed with malignant glioma.** (DeLorenze et al. 2010). === *For patients diagnosed with Grade II and Grade III histology, moderate (915.8-2118.3 mcg) intake of fat-soluble lycopene was associated with poorer survival* when compared to low intake (0.0-914.8 mcg), for self-reported cases only. *In Grade IV patients, moderate/high intake of cryptoxanthin and high intake of secoisolariciresinol were associated with poorer survival among all cases.* Among **Grade II patients,** moderate intake of water-soluble folate was associated with greater survival for all cases; *high intake of vitamin C and genistein and the highest level of the antioxidant index were associated with poorer survival for all cases.*

Total number of participants in the above studies totals 3,129,215. Note: one study on green tea did not give the number of participants but it did include 13 epidemiologic studies.

<u>**SUMMARY**</u> **of antioxidants causing various cancers include the following: lung cancer, breast cancer, prostate cancer, stomach cancer, esophageal cancer, head and neck cancer, bladder cancer, melanoma, adenomas, oral premalignant lesions.**

Antioxidants increased the mortality in the following cancer patients (2 studies):

- 1) increase overall mortality in gastrointestinal cancer patients,

- 2) esophageal cancer deaths increased 14% in patients, (2 studies)

The following are studies in which the antioxidant vitamins either had no effect or had a harmful effect: (part taken from *Death in small doses?* and part from *Antioxidant Overkill*). Some new studies were also added.

Vitamin E and N-acetylcysteine (NAC) Can Fuel the Growth of Lung Cancers

Now, **a team of Swedish scientists has shown that two antioxidants—vitamin E and N-acetylcysteine (NAC)—can fuel the growth of lung cancers in mice.**

<u>Martin Bergo's</u> team at the University of Gothenburg showed that **antioxidants neutralize ROS in tumors as well as healthy cells. "If we give extra antioxidants in the diet, we're helping the tumor to reduce radicals that would otherwise block its growth."**

"They might have a small undiagnosed tumor, and no one knows the frequency of those," he added. **"There's a possibility that antioxidants would speed up the growth of those**

tumors." This word of caution is especially relevant to people with COPD, who often take large amounts of NAC to relieve the build-up of mucus in their airways.

"A warning seems appropriate for everyone who has been seduced to use antioxidants or vitamins on a regular basis, as a preventive measure," the University of Syndey's Nico van Zandwijk told *The Scientist* in an e-mail.

These results fit with those from a long line of human clinical trials, in which antioxidants failed to prevent disease or made things worse. The first of these was published in the New England Journal of Medicine in 1994, and showed that male smokers who took beta-carotene supplements were more likely to develop and die of lung cancer than those who did not. Other trials found similar results for other antioxidants and other cancers, and some of those studies were even stopped early.

In 2012, the Cochrane Collaboration analyzed the results of 78 earlier trials and, based on the most careful of them, concluded that **people who took antioxidant supplements (including both healthy people and those with chronic diseases) were more likely to die prematurely than those who did not.**

Few studies had looked at the reasons behind these **seemingly paradoxical effects**. Bergo's team, led by graduate student Volkan Sayin, began by feeding NAC and vitamin E to mice with early lung cancers, at doses comparable to those in human multivitamin pills. **The mice that ingested the antioxidants developed tumors that were three times bigger, and they died twice as fast.**

Sayin then showed that tumors normally have lower levels of ROS than normal tissues. The antioxidants reduced these levels even further, protecting the tumors from DNA damage. They also dramatically reduced the activity of p53—a guardian protein that prevents cancer by detecting damaged DNA and putting the brakes on cell division.

By lifting p53's suppression, the antioxidants allowed the cancer cells to grow and divide faster than usual. Indeed, when the team abolished p53 entirely, neither NAC nor vitamin E affected the growth of the lung tumors. "P53 is normally inactivated in late-stage lung cancer, so what we're doing is speeding up the progression of malignancy," said Bergo.

"This is an extremely striking observation, but not surprising given the rather disappointing outcomes of patients at risk for developing lung cancer who had been treated with various antioxidants," David Tuveson from Cold Spring Harbor Laboratory, who was not involved in the work, said in an e-mail. **"We should now consider whether people consuming high doses of antioxidants are ironically promoting cancers that they seek to prevent."**

Although Bergo's team focused on mice in this study, **the scientists found the same mechanisms at work in human cells.** V.I. Sayin et al., "Antioxidants accelerate lung cancer progression in mice," *Science Translational Medicine*, 6: 221ra15, 2014.

The National Cancer Institute already advises cancer patients that antioxidant supplements "should be used with caution."

In essence, "antioxidants allow cancer cells to escape cells' own defense system" against tumors, biologist Per Lindahl of Sweden's University of Gothenburg and a co-author of the study told reporters. **That lets existing tumors, even those too small to be detected, proliferate uncontrollably.**

The findings imply that "taking extra antioxidants might be harmful and could speed up the growth of (any) tumors," said **biologist and co-author Martin Bergo of Gothenburg, adding, "If I had a patient with lung cancer, I would not recommend they take an antioxidant."** January 29, 2014

The Simplicity of my EMOD Induced Apoptotic Approach

In studying the full complexity of cancer, I have found that an important underlying simplicity has emerged, i.e., the undeniable crucial role of EMODs.

Cancer is mind-bogglingly complicated. Every tumor contains groups of cells with different genetic mutations, and even within genetically identical groups, genes are expressed at different levels. Those cells interact with the body's connective fibers, blood vessels, immune cells, and so on. They're subject to varied forces and pressures. And that's just one tumor. Each kind of breast cancer tumor differs substantially from the others, and even more from tumors produced by lung cancer, brain cancer, pancreatic cancer, and so on.

My work extracts fundamental simplicity from a seemingly hopeless mess of complexity.

That quest for simplicity, however, comes with an important caveat. You may have heard this aphorism, often attributed to Albert Einstein: "Everything should be made as simple as possible, but not simpler." Einstein actually said, "It can scarcely be denied that the supreme goal of all theory is to make the irreducible basic elements as simple and as few as possible without having to surrender the adequate representation of a single datum of experience."

Cancer has resisted a decades-long assault that has cost more than $100 billion.

Metastasis is responsible for nine in 10 cancer deaths. Surgeons can often cut out a primary tumor, but there's no way to find and excise dozens or hundreds of far-flung growths, many of which don't show up on scans.

Yet, my EMOD induced apoptosis will kill both the primary cancer and the metastatic sites.

March 07, 2018
The State of Cancer: Are We Close to a Cure?
Healthline/Medical News Today

Cancer is the leading cause of death across the globe. For years now, researchers have led meticulous studies focused on how to stop this deadly disease in its tracks. How close are we to finding more effective treatments?

The World Health Organization (WHO) note that, worldwide, nearly 1 in 6 deaths are due to cancer.

In the United States alone, the National Cancer Institute (NCI) estimated 1,688,780 new cancer cases and 600,920 cancer-related deaths in 2017.

Currently, the most common types of cancer treatment are chemotherapy, radiotherapy, tumor surgery, and—in the case prostate cancer and breast cancer—hormonal therapy.

However, other types of treatment are beginning to pick up steam: therapies that—on their own or in combination with other treatments—are meant to help defeat cancer more efficiently and, ideally, have fewer side effects.

Innovations in cancer treatment aim to address a set of issues that will typically face health-care providers and patients, including aggressive treatment accompanied by unwanted side effects, tumor recurrence after treatment, surgery, or both, and aggressive cancers that are resilient to widely utilized treatments.

Boosting the Immune System's 'Arsenal'

One type of therapy that has attracted a lot of attention recently is immunotherapy, which aims to reinforce our own bodies' existing arsenal against foreign bodies and harmful cells: our immune system's response to the spread of cancer tumors.

But many types of cancer cell are so dangerous because they have ways of "duping" the immune system—either into ignoring them altogether or else into giving them a "helping hand."

Therefore, some types of aggressive cancer are able to spread more easily and become resistant to chemotherapy or radiotherapy.

However, thanks to *in vitro* and *in vivo* experiments, researchers are now learning how they might be able to "deactivate" the cancer cells' protective systems. A study published in 2017 in *Nature Immunology* found that macrophages, or white blood cells, that are normally tasked with "eating up" cellular debris and other harmful foreign "objects" failed to obliterate the super-aggressive cancer cells.

That was because, in their interaction with the cancer cells, the macrophages read not one but two signals meant to repel their "cleansing" action.

This knowledge, however, also showed the scientists the way forward: by blocking the two relevant signaling pathways, they re-enabled the white blood cells to do their work.

Therapeutic Viruses and Innovative 'Vaccines'

A surprising weapon in the fight against cancer could be therapeutic viruses, as revealed by a team from the United Kingdom earlier this year. In their experiments, they managed to use a reovirus to attack brain cancer cells while leaving healthy cells alone.

"This is the first time it has been shown that a therapeutic virus is able to pass through the brain-blood barrier," explained the study authors, which "opens up the possibility [that] this type of immunotherapy could be used to treat more people with aggressive brain cancers."

Another area for improvement in immunotherapy is "dendritic vaccines," a strategy wherein dendritic cells (which play a key role in the body's immune response) are collected from a person's body, "armed" with tumor-specific antigens—which will teach them to "hunt" and destroy relevant cancer cells—and injected back into the body to boost the immune system.

In a new study, researchers in Switzerland identified a way to improve the action of these dendritic vaccines by creating artificial receptors able to recognize and "abduct" tiny vesicles that have been linked to cancer tumors' spread in the body.

By attaching these artificial receptors to the dendritic cells in the "vaccines," the therapeutic cells are enabled to recognize harmful cancer cells with more accuracy.

Importantly, recent studies have shown that immunotherapy may work best if delivered in tandem with chemotherapy—specifically, if the chemotherapy drugs are delivered first, and they are followed up with immunotherapy.

But this approach does have some pitfalls; it is difficult to control the effects of this combined method, so sometimes, healthy tissue may be attacked alongside cancer tumors.

However, scientists from two institutions in North Carolina have developed a substance that, once injected into the body, becomes gel-like: a "bioresponsive scaffold system." The scaffold can hold both chemotherapy and immunotherapy drugs at once, releasing them systematically into primary tumors.

This method allows for a better control of both therapies, ensuring that the drugs act on the targeted tumor alone.

The Nanoparticle Revolution

Speaking of specially developed tools for delivering drugs straight to the tumor and hunting down micro tumors with accuracy and efficiency, the past couple of years have seen a "boom" in nanotechnology and nanoparticle developments for cancer treatments.

Nanoparticles are microscopic particles that have garnered so much attention in clinical research, among other fields, because they bring us the chance to develop precise, less invasive methods of tackling disease.

Vitally, they can target cancer cells or cancer tumors without harming healthy cells in the surrounding environment.

Some nanoparticles have now been created to provide very focused hyperthermic treatment, which is a type of therapy that uses hot temperatures to make cancer tumors shrink.

Last year, scientists from China and the UK managed to come up with a type of "self-regulating" nanoparticle that was able to expose tumors to heat while avoiding contact with healthy tissue.

"This could potentially be a game-changer in the way we treat people who have cancer," said one of the researchers in charge of this project.

These tiny vehicles can also be used to target cancer stem-like cells, which are undifferentiated cells that have been linked to the resilience of certain types of cancer in the face of traditional treatments such as chemotherapy.

Thus, nanoparticles can be "loaded" with drugs and set to "hunt down" cancer stem cells to prevent the growth or recurrence of tumors. Scientists have experimented with drug-filled nanoparticles in the treatment of various types of cancer, including breast and endometrial cancer.

No less importantly, minuscule vehicles called "nanoprobes" can be used to detect the presence of micrometastases, which are secondary tumors so tiny that they cannot be seen using traditional methods.

Dr. Steven K. Libutti, director of the Rutgers Cancer Institute of New Jersey in New Brunswick, calls micrometastases "the Achilles' heel of surgical management for cancer" and argues that nanoprobes "go a long way to solving [such] problems."

Tumor 'Starvation' Strategies

Another type of strategy that researchers have been investigating of late is that of "starving" tumors of the nutrients they need to grow and spread. This, scientists point out, could be a saving grace in the case of aggressive, resilient cancers that cannot effectively be eradicated otherwise.

Three different studies—whose results were all published—looked at ways of cutting off cancers' nutritional supplies.

One of these studies looked at ways of stopping glutamine, a naturally occurring amino acid, from feeding cancer cells.

Certain cancers, such as breast, lung, and colon, are known to use this amino acid to support their growth.

By blocking cancer cells' access to glutamine, the researchers managed to maximize the impact of oxidative stress, a process that eventually induces cell death, on these cells.

Some aggressive types of breast cancer may be halted by stopping the cells from "feeding" on a particular enzyme that helps them to produce the energy they need to thrive.

Another way of depleting cancer cells of energy is by blocking their access to vitamin B-2, as researchers from the University of Salford in the UK have observed.

As one study author says, "This is hopefully the beginning of an alternative approach to halting cancer stem cells." This strategy could help individuals receiving cancer treatment to avoid the toxic side effects of chemotherapy.

Cancer Treatments and Epigenetics

Epigenetics refers to the changes caused in our bodies by alterations in gene expression, which dictate whether certain characteristics appear or if certain "actions" are affected at a biological level.

According to research that addressed the impact of such changes, many cancers, as well as the behaviors of cancer cells, are determined by epigenetic factors.

"Recent advances in the field of epigenetics have shown that human cancer cells harbor global epigenetic abnormalities, in addition to numerous genetic alterations."

"These genetic and epigenetic alterations interact at all stages of cancer development, working together to promote cancer progression."

Thus, it is crucial for specialists to understand when and where to intervene and the expression of which genes they may need to switch on or off, depending on their role in the development of cancer.

One study, for instance, found that the gene responsible for the advent of Huntington's disease produces a set of molecules whose action may actually prevent cancer from occurring.

Now, the researchers' challenge is to channel the therapeutic potential of this process without triggering Huntington's disease. However, the scientists are hopeful.

"We believe a short-term treatment cancer therapy for a few weeks might be possible," says the study's senior author.

Another recent study was able to establish that estrogen-receptor positive breast cancers that become resistant to chemotherapy gain their resilience through genetic mutations that "confer a metastatic advantage to the tumor."

But this knowledge also gave researchers the "break" that they needed to come up with an improved treatment for such stubborn tumors: a combination therapy that delivers the chemotherapeutic drug fulvestrant alongside an experimental enzyme inhibitor.

What Does This all Mean?

Cancer research is running at full speed, taking advantage of all the technological advances that science has achieved over recent years. But what does that mean in terms of coming up with a cure for cancer?

Whether or not there will ever be a cure for all cancer types is currently a matter of strong debate; although promising studies are published and covered by the media almost every day, cancer types vary immensely.

This makes it very difficult to say that an approach that works for one type will be adaptable to all.

Also, while there is much emerging research promising more effective treatments, most of these projects are still in their early stages, having conducted *in vitro* and *in vivo* experiments. Some potential treatments still have a long way to go before clinical trials in human patients.

Still, that doesn't mean we should lose all hope. Some researchers explain that these efforts should make us optimistic; while we may not be at the stage where we can claim that cancer can easily be eradicated, our furthered knowledge and ever more precise tools keep us ahead of the game and improve our odds in the fight against this disease.

The latest estimates of worldwide figures suggest that there were 14.1 million new cases of cancer in 2012, and that this number is expected to climb to 24 million by 2035.

In the United States — where cancer is the second most common cause of death — the American Cancer Society (ACS) estimate that there will be around 1.7 million newly diagnosed cases of cancer, and more than 609,000 deaths to the disease, in 2018.

According to the ACS, at least 42 percent of newly diagnosed cases of cancer are preventable. These include 19 percent in which smoking is the main cause and 18 percent that result from a combination of factors, including "poor nutrition."

Observational studies are not designed to prove cause and effect — but they can offer insights into links between variables such as diet and disease.

What Are Trends in Cancer Mortality Rates?

Authors: News Author: Pam Harrison; CME Author: Charles P. Vega, MD

CME / ABIM MOC / CE Released: 3/2/2018

Clinical Context

One problem with statistics gleaned from national databases is that they tend to lag at least 2 years behind the present. To address this issue, the authors of the current study used national data from 2000 to 2014 to estimate not only the number of cases of invasive cancer expected in 2018 but also the rate of cancer mortality.

They estimate that there will be 1,735,350 new cases of cancer in the United States in 2018, which equates to an average of more than 4700 cases per day.

There is little change expected for the most common types of cancer, which are prostate, lung, and colorectal among men. Together, these tumors account for 42% of all cancers among men; nearly 20% of new cancers among men involve the prostate.

Women have a lower lifetime risk for cancer compared with men, and interestingly, height may influence this disparity. **The types of cancer expected to continue to be most common among women are breast, lung, and colorectal**. These 3 tumors account for half of all cancer cases among women, with breast cancer itself responsible for 30% of new cancers.

More than 600,000 Americans are expected to die of cancer in 2018. Cancers of the lung, breast, colorectum, and prostate will account for 45% of these deaths.

Study Synopsis and Perspective

Mortality from most cancers continues to decline across the United States, although there is still concern that liver cancer continues to "increase rapidly," according to new figures released by the American Cancer Society.

The figures come from the annual report, *Cancer Statistics 2018*, which was published online January 4, 2018 in *CA: A Cancer Journal for Clinicians*.

From a peak death rate in 1991, mortality from all cancers had declined by 26% by 2015.

This means that there are approximately 2.4 million fewer deaths from cancer now than what would have occurred had peak mortality rates remained at 1991 levels, the authors note.

"The decline in cancer mortality over the past 2 decades is primarily the result of steady reductions in smoking and advances in early detection and treatment," they comment.

The largest fall in cancer deaths has been for lung cancer deaths in men, which decreased by 45% between 1990 to 2015

The smoking connection was emphasized by Otis Brawley, MD, chief medical officer of the American Cancer Society. "A decline in consumption of cigarettes is credited with being the most important factor in the drop-in cancer death rates," he said in a statement.

"Strikingly, though, tobacco remains by far the leading cause of cancer deaths today, responsible for nearly 3 in 10 cancer deaths," he added.

Study Synopsis and Perspective

The researchers also found that during the last decade, the overall incidence of cancer declined by about 2% a year in men. This reflects large, continuing declines in the rates of both lung and colorectal cancer, as well as a decrease in the incidence of prostate cancer.

Indeed, the incidence of lung cancer in men is dropping about twice as rapidly in men as in women, the authors note, a reflection of the fact that men took up smoking earlier than women, as well as the fact that more men are quitting smoking than women.

Rates of colorectal cancer are generally similar in men and women and have similarly declined by about 2% to 3% a year from 2005 through 2014.

This reflects more widespread screening by colonoscopy among older adults throughout the United States, the authors suggest.

In contrast, "the overall cancer incidence rate in women has remained generally stable over the past few decades because declines in lung and colorectal cancers have been offset by increasing or stable rates for breast, uterine corpus and thyroid cancers and for melanoma," the researchers observe.

Liver Cancer Continues to "Increase Rapidly"

Trends in liver cancer seen between 2010 and 2014 are interesting because they vary by age and sex.

Overall, the incidence of liver cancer continues to "increase rapidly" in women, although not in men. However, rates are increasing in both sexes in those younger than 40 years; rates remained stable or declined in men and women aged 40 to 59 years.

For individuals aged 60 to 69 years, the incidence of liver cancer increased by 8% each year between 2010 and 2014 and by 3% for those 70 years of age and older, the researchers note.

This pattern likely reflects detection rates of hepatitis C infection (HCV), as well as new antiviral therapies that reduce the risk for liver cancer, they observe.

In contrast, there is a "worrisome 2-fold increase in HCV infections from 2010 to 2014...driven by individuals aged 20-39 years as a consequence of the opioid epidemic," investigators state.

As for melanoma, the rapid rise in the incidence of this skin malignancy that occurred between 2005 and 2014 appears to be slowing down, especially among younger individuals.

Reflecting changes in practice guidelines, thyroid cancer rates have also begun to stabilize in recent years.

Survival Rates

From a peak death rate in 1991, mortality from all cancers had declined by 26% by 2015, but the fall in cancer death rates was larger in men than in women (32% vs 23%), the authors report.

Large declines across the decades have been recorded for all 4 of the main cancer types in both men and women:

- 39% decline in breast cancer in women from 1989 to 2015,
- 52% decline in prostate cancer from 1993 to 2015,
- 52% decline in colorectal cancer from 1970 to 2015, and
- 19% decline in lung cancer in women from 2002 to 2015.

"For all cancers combined, the 5-year relative survival rate is 68% in whites and 61% in blacks," researchers continue.

Prostate cancer is associated with the highest survival rates, at 99% at 5 years, followed by melanoma, at 92%, and breast cancer in women, at 90%.

The lowest survival rates are seen with pancreatic cancer, at 8% at 5 years, followed by lung and liver cancer, both at 18%.

"Survival is lower for black than for white patients for every cancer type...except cancers of the kidney and pancreas," the researchers note.

As has been widely reported elsewhere, black patients are more likely than white patients to present with more advanced cancer. After adjusting for all confounders, including stage of diagnosis, "the relative risk of death after a cancer diagnosis is 33% higher in black patients than in white patients," the researchers observe.

Among all ages combined, death rates from cancer were 14% higher in non-Hispanic blacks than in non-Hispanic whites.

These rates are 33% lower relative to mortality rates in 1993, and the mortality gap between blacks and whites is now even lower, at 7%, in patients 65 years of age and older, likely because of better access to healthcare offered by Medicare.

In fact, mortality rates from cancer are lower among blacks in Massachusetts and New York than they are for white patients.

Massachusetts now offers almost universal healthcare coverage, which may have helped eliminate racial disparities in cancer mortality rates, the authors note.

Cancer is the second most common cause of death in children aged 14 years and younger. The researchers estimate that 10,590 children will be diagnosed with cancer in 2018, and that 1180 will die from it.

Leukemia is the most common childhood cancer, followed closely by brain tumors and other nervous system tumors.

Adolescents are most susceptible to brain tumors and other nervous system tumors, followed closely by lymphoma.

CA Cancer J Clin. Published January 4, 2018.[1]

Study Highlights

- Researchers used multiple national databases to evaluate trends in the prevalence and mortality of cancer.
- The incidence of cancer among men has declined by about 2% per year during the last decade, with a more rapid decline in the past several years. Cancer of the lung and

colorectum continues to be less common, but the incidence of prostate cancer has declined approximately 10% annually from 2010 to 2014.

- In contrast, the incidence of cancer has remained stable among women. Reductions in lung and colorectal cancer have been offset by higher numbers of women with breast, uterine, and thyroid cancer, as well as melanoma.
- Rates of colonoscopy among US adults older than 50 years have evolved from 21% in 2000 to 60% in 2015, which has helped reduce the incidence of colorectal cancer.
- The incidence of liver cancer has leveled off among men but continues to climb among women. Powerful new drugs against HCV should improve the risk for liver cancer over time, but only 14% of baby boomers recommended for routine HCV screening have gotten the test done.
- The incidence of melanoma has increased for more than a decade, but the rate of growth has slowed.
- The average 5-year survival rate for cancer is 68% among whites and 61% among blacks. The rate of cancer mortality declined by approximately 1.5% per year from 2006 to 2015 among women and men.
- Cancers with the highest 5-year survival rates include prostate (99%), melanoma (92%), and breast cancer (90%). Cancers with the worst respective survival rates include tumors of the pancreas (8%), lung (18%), and liver (18%).
- After adjustment for age, sex, and tumor stage, the risk for cancer mortality is 33% higher among black patients compared with white patients. The disparity in cancer mortality is particularly pronounced among individuals younger than 65 years.
- The risk for death associated with hematologic and lymphoid cancers has declined dramatically, thanks to better treatment options. The average 5-year survival rate for chronic myeloid leukemia has increased from 22% during the mid-1970s to 68% from 2007 to 2013.
- In contrast, treatment advances have been less effective for tumors of the pancreas and lung.
- Rates of lung cancer screening among appropriate patients remain below 5%.
- Cancer is the second-leading cause of death, after heart disease, but it was the only 1 of the top 10 causes of mortality to decline in incidence between 2014 and 2015.
- Cancer is the leading cause of death among Asian Americans and Hispanics in the United States.
- Cancer is second only to accidents as a cause of death among children between the ages of 1 and 14 years. The projections for cancer incidence and mortality in 2018 include 10,590 and 1180 children, respectively.
- Leukemia accounts for 29% of pediatric cancer cases, and tumors of the central nervous system account for another 26%.

Clinical Implications

- The most common types of cancer expected to affect women and men in 2018 remain breast and prostate cancer, respectively.
- The incidence of cancer has declined among US men in the past decade, but it has remained stable among women. Cancer mortality rates have been declining steadily for a decade for women and men.

- Implications for the Healthcare Team: Screening rates for lung cancer among appropriate patients remain unacceptable, and the healthcare team should develop protocols to test older adults for HCV as well.

Study: Those High-Priced Antioxidants May Be Killing You

20Feb 2008 - by Stewart Lawrence

BEIJING — Fear of mortality is one reason Americans spend so much on "antioxidant" products, including Vitamin C supplements and beta-carotene, which promise a longer healthier life. **According to the National Institutes of Health, more than half of adults in the U.S. consume some kind of antioxidant product, spending $37 billion each year**.

But a study conducted in China – where aging is akin to a national obsession these days – claims that antioxidants don't work as billed. The study is published in the journal *Redox Biology*.

A new study finds that antioxidant supplements may be more harmful to the human body than believed.

Rather than extending longevity, researchers say they trigger a stress reaction which causes the body to age more rapidly.

In other words, those expensive life-enhancers may actually be killing you.

Researchers discovered the relationship by studying how oxidants affected worms and human cells at various stages of development. **Oxidants, it turned out, had no measurable impact on aging.**

But introducing antioxidants disturbed the mechanism in cells that resists aging and as a result, the cells began aging more rapidly — "unnaturally fast," said Chen Chang, the lead scientist on the project at the Institute of Biophysics, part of the Chinese Academy of Sciences in Beijing.

Chang and her team are especially concerned about their findings because of the widespread use of antioxidants among Chinese youth.

"More and more 'white collar workers' in their 20s are taking pills containing antioxidants such as Vitamin C and tea polyphenols. They must stop," **Chang told the South China Morning Post.**

But older people are also more likely to die faster due to supplements, they found. Chang recommended that anti-oxidant use in China and elsewhere be curtailed.

The Chinese study appears to confirm the results of a 2008 "meta-review" of 405 studies comparing the longevity of people that consumed various kinds of vitamin supplements with those that took only a placebo.

That review, conducted by top medical researchers in Denmark, found no significant differences in the mortality rates of the two groups.

So far, none of these new studies has put a dent in the multi-billion-dollar vitamin and supplement industry, which continues to promote the presumed anti-aging benefits of its products.

And there is little evidence that consumers are paying attention, either – or if they are, that they care.

But the evidence against antioxidants is mounting.

Another medical study conducted last year found that overuse of Vitamin D supplements was associated with a higher risk of falls in men and in women 70 years and older.

Apparently, Nature is sending a message: Stop trying to reset your biological clock.

Dietary Antioxidants and Human Cancer

(Borek, Carmia, 2017)

Human observational studies suggest that high intake of antioxidant-rich foods, notably fruits, vegetables, and grains, is inversely related to cancer risk. However, **prospective human studies that began to accrue data in the 1990s did not always confirm this relationship. Randomized, controlled clinical trials using specific antioxidant supplementation have also produced mixed results**. Antioxidant supplementation during cancer treatment remains controversial. Radiotherapy and certain chemotherapeutic agents rely on EMODs to destroy cancer cells. Although antioxidant supplementation may help protect normal cells from EMODs damage and may have palliative effects during cancer treatment, studies suggest that cancer cells may also be protected from ROS damage, thereby reducing treatment efficacy and patient survival. This article reviews evidence for the impact of antioxidant supplementation and antioxidant-rich diets on cancer risk and mortality. It also outlines some of the factors that may have contributed to the conflicting outcomes reported.

INTRODUCTION

Cancer is the second leading cause of death in the United States. The American Cancer Society estimates that **in 2017 there will be 1,688,780 new cancer cases and 600,920 cancer deaths**. (American Cancer Society. Cancer Statistics Center: 2017 estimates. Atlanta, GA: American Cancer Society. Available from: https://cancerstatisticscenter.cancer.org/#/)

Cancer cells are distinguished by their limitless replicative potential, resistance to growth inhibition, metastatic spread, and evasion of apoptosis. On the basis of observational and experimental studies, diets rich in fruits and vegetables have unsuccessfully ranked highly as a strategy to lower cancer risk because they contain antioxidants capable of destroying ROS.

Prospective studies since the 1990s have introduced uncertainty into the perception of the cancer-preventive efficacy of these plant-rich diets. Radiotherapy and certain chemotherapies use EMODs to destroy neoplastic cells. Dietary antioxidant supplements such as vitamins C and E have been considered as adjuvants in cancer therapy because they have been shown to mitigate the adverse effects of radiation in some patients. However, **they have also been shown to protect cancer cells, thereby reducing treatment efficacy.** (Borek, 2004) (Lawenda et al, 2008) (Borek C, Pardo F., 2002) (Sigounas G, Anagnostu A, Steiner M., 1997) (Bairati I, Meyer F, Jobin E, et al., 2006) (Ma Y, Chapman J, Levine M, et al., 2014)

ANTIOXIDANTS IN CANCER DEVELOPMENT AND TREATMENT

EMODs are by-products of normal cellular metabolism, but their levels can increase markedly during inflammation and exposure to certain exogenous factors such as ionizing radiation; nitrogen oxide pollutants; some chemical carcinogens; tobacco smoke; and **particular drugs, such as acetaminophen.**

ANTIOXIDANTS DURING CANCER TREATMENT

It is difficult to come to a consensus about the effects of antioxidant supplementation during treatment because of the variability in the studies in which researchers have investigated this question. **These include variations in study design, eligibility criteria, intervention protocols, malignancy type, treatment modality, antioxidant form and dose, timing of observation and intervention, and statistical power.**

It also has to be borne in mind that dietary antioxidants comprise a broad range of chemical classes, including carotenoids, polyphenols, tocopherols, and others that have a wide array of biological activities. In addition, although antioxidant supplements share the ability to quench EMODs, their clinical action and potency can vary by type, bioavailability, dose, duration, and route of administration.

Currently, there is sufficient evidence from trials to determine that excessive antioxidant supplementation taken during treatment for protection of normal cells from EMOD damage provides similar protection to cancer cells and thus diminish treatment effect.

In vitro studies indicate that antioxidants kill tumor cells by apoptosis while sparing normal cells and appears to be replicated in clinical trials.

Researchers in a few, randomized controlled trials found that **participants who took antioxidant supplements had worse outcomes, especially if they were smokers.**

In one study, 540 patients with head and neck cancer received α-tocopherol (400 IU/day) with or without β-carotene (30 mg/day) versus placebo, concurrent with radiation therapy. The median duration of radiation therapy was 43 days, and the median duration of antioxidant supplementation was 3.1 years. During the follow-up (median, 6.5 years), 179 deaths were recorded for the supplement group and 77 for the placebo group. **All-cause mortality was significantly increased in patients who had received α-tocopherol alone and those who received β-carotene supplements in addition to α-tocopherol.** The addition of β-carotene did not markedly affect the mortality rate.

ANTIOXIDANT FOODS AND CANCER PREVENTION

The potential role of antioxidant-rich fruit and vegetables in helping prevent the onset of diseases, including cancer, has been the subject of basic research and observational and intervention studies. Observational studies showed that the risk of certain cancers is inversely related to consumption of total fruits and total vegetables that contain essential antioxidant micronutrients and phytochemicals.

The reported results were based mainly on case–control studies, which are notoriously susceptible to recall bias when it comes to reporting on past diet. In the late 1990s, large prospective cohort studies on diet and cancer began to accrue data that did not always confirm the inverse associations reported in the observational studies.

In a large prospective study (the EPIC study), investigators looked at a European cohort of 142,605 men and 335,873 women who developed approximately 30,000 cancers at all sites combined over nearly 9 years of follow-up. The investigators found a weak but statistically significant inverse association – a 4% lower incidence of all cancers combined for an increment of 200 g of total fruits and vegetables per day, which corresponds to about two extra servings per day.

Regarding the results for specific cancers, **analysis of the data showed an inverse relationship between cancers of the upper gastrointestinal tract and fruit intake but not vegetable intake**. The risk of colorectal cancer was inversely associated with intake of total fruits and vegetables and total fiber. Lung cancer in smokers was inversely associated with fruit intake but not with vegetable intake. **No significant association was found between the intake of total fruits, vegetables, and fiber and observed cancers of the stomach, biliary tract, pancreas, cervix, endometrium, prostate, kidney, bladder, or lymphoma**.

A prospective study performed within the NIH-AARP Diet and Health Study showed a different pattern of results. The investigators identified 15,792 cancer cases in 195,229 women and 35,071 cancer cases in 288,109 men, and they estimated relative risks associated with the highest compared with the lowest quintile of fruit and vegetable intakes. The complex results highlight the difficulty in reaching conclusions. **Total cancer among women and men was not associated with the intake of fruits or vegetables.**

However, analysis of risk for cancers at specific sites revealed that men had a lower risk of pancreatic cancer with fruit intake, colorectal cancer and lung cancer with vegetable intake, and **an increased risk for prostate and thyroid cancer with vegetable intake.**

Fruit intake lowered the risk of breast cancer in women, but an intake of vegetables increased breast cancer risk.

The conclusion of the investigators based on total cancer was that the results did not support an association between fruits and vegetables and total cancer in men and women.

A particular food component may provide a more selective benefit than others. In a case–control study, the intake of allium vegetables (onions, leeks, garlic) high in organosulfur compounds and other antioxidants selectively reduced gastric cancer. (Boffeta P, Wichmann J, Ferrari P, et al., 2010)

Authors of a systematic review of prospective studies on citrus fruit and cancer found that with each 100 g/day increased intake of citrus fruits, there was a marginally significant decreased risk of esophageal cancer (six studies), squamous cell carcinoma (three studies), and esophageal adenocarcinoma (three studies). A nonsignificant inverse relationship was found for gastric cardia cancer but not for gastric noncardiac cancer (four studies).

Evidence for fruits and vegetables lowering the risk of hormonally regulated cancers is inconsistent. Nevertheless, some studies have reported a protective effect.

In case–control studies of Asian women, green tea, which is rich in antioxidant polyphenols, significantly reduced the risk of breast cancer. The consumption of cooked tomatoes and tomato products, which are rich in the carotenoid lycopene, a powerful antioxidant, was linked to lower incidences of a number of cancers, most strongly prostate cancer.

A weak association between intake of fruits and vegetables and cancer risk may still represent a true protective effect. Even if the evidence for the cancer-preventive effects is not as strong as previously thought, accumulating data support benefits for preventive effects against cardiovascular disease. For example, men and women who showed no inverse relationship between total cancer rates and intake of fruits and vegetables in the EPIC study showed a 30% reduction in the incidence of coronary heart disease or stroke when consuming five or more servings per day, compared with those eating less than 1.5 servings per day. Thus, recommendations to increase the intake of fruits and vegetables have a somewhat sound basis.

Differences in the outcomes of studies on the protective effects of antioxidant-rich foods may be due in part to differences in genetics, age, sex, and lifestyle in study participants and differences in the food itself. Fruits and vegetables vary widely in their composition and their antioxidant capability. Antioxidant levels in any particular plant food depend on growing conditions, such as soil composition, moisture, and temperature. Variations in the handling and storage of food after harvest, as well as differences in food preparation methods, also affect antioxidant composition. In addition, several endogenous enzymes that scavenge EMODs, such as superoxide dismutase, catalase, and peroxidases, depend on dietary micronutrients for their activity. Copper and zinc act as coenzymes for superoxide dismutase; iron is required for catalase; and selenium, a dietary trace element, is a component of the enzyme glutathione peroxidase that breaks down the ROS peroxide.

Thus, any deficiency of these cofactors in study participants, or any depletion in vegetables and fruits used as interventions, might also contribute to the inconsistent findings on the protective effects of antioxidant-rich foods.

ANTIOXIDANT SUPPLEMENTS AND CANCER PREVENTION

Observational studies, including case–control and cohort studies, on the effect of antioxidant supplements in cancer prevention have shown mixed results. (Patterson RE, White E, Kristal AR, et al., 1997)

Because observational studies cannot control for a variety of biases that may influence outcome, each study must be critically evaluated. Remember that none of these observational studies "prove" anything.

By contrast, randomized, controlled clinical trials lack most of the limiting biases encountered in observational studies. Such intervention studies are considered more reliable evidence of benefits or harm. Randomized controlled trials on the role of dietary antioxidant supplements in cancer prevention have looked at a small number of essential dietary antioxidants, including β-carotene, vitamin A, vitamin E, vitamin C, and selenium, given mostly in large doses and lasting relatively short periods of time. **Fraught by limitations in the studies, results have been mixed**.

The researchers in the Alpha-Tocopherol, Beta-Carotene Cancer Prevention Trial (ATBC), which was done in Finland, investigated whether daily intake of vitamin E (α-tocopherol) and β-carotene over 5–8 years would lower lung cancer incidence in male smokers. **The results showed that β-carotene at 20 mg/day increased the incidence of lung cancer.**

The researchers in the Carotene and Retinol Efficacy Trial (CARET) investigated the effect of daily β-carotene (30 mg) and retinyl palmitate (25,000 IU) on the incidence of lung cancer and other cancers, as well as on death, in participants who were at high risk for lung cancer because of a history of smoking or asbestos exposure. **CARET began in 1986 and was stopped ahead of schedule in January 1996 because participants who were randomly assigned to receive the intervention showed a 28% increase in incidence of lung cancer, a 17% increase in incidence of death, and a higher rate of cardiovascular disease mortality than participants in the placebo group.** (Omenn GS, Goodman GE, Thorquist MD, et al., 1996)

A report in 2004 showed that the adverse effects persisted for up to 6 years after stopping supplementation, though the increased risk of lung cancer and all-cause mortality observed earlier were no longer statistically significant compared with placebo. (Goodman GE, Thornquist MD, Balmes J, et al., 2004)

The investigators in the Physicians' Health Study-I examined the effects of β-carotene at 50 mg every other day for 12 years on cancer mortality or all-cause mortality in healthy male physicians. The results reported in 1996 showed no effects in smokers and nonsmokers. (Hennekens CH, Buring JE, Manson JE, et al., 1996)

These results differ from the CARET and ATCB trials, where cancer incidence was higher in smokers taking β-carotene and no clear explanation was provided.

The researchers in the Physicians' Health Study-II investigated whether 400 IU of vitamin E taken every other day, 500 mg of vitamin C taken every day, or a combination of both would have an effect on cancer incidence in healthy male physicians aged 50 years and older. **The study was carried out over 7.6 years, and the results, reported in 2009, showed that supplementation with either antioxidant or both did not reduce cancer incidence.** (Gaziano JM, Glynn RJ, Christen WG, et al., 2009)

The investigators in the Women's Health Study examined the effects of β-carotene (50 mg every other day), vitamin E (600 IU every other day), and aspirin (100 mg every other day) on cancer incidence and cardiovascular disease in U.S. women aged 45 years and older. **The results, reported in 1999, showed no benefit or harm associated with supplementation.** (Lee IM, Cook NR, Manson JE, et al., 1999)

A report in 2005 showed similar no benefit results for vitamin E. (Lee IM, Cook NR, Graziano JM, et al., 2005)

The role of selenium in cancer prevention has been of interest for some time. Experimental studies have shown selenium to have protective effects, and it is known that regions low in selenium have higher rates of cancer. A large multicenter clinical trial showed that subjects with a history of skin cancer who were supplemented with 200 mg of selenium for a mean of 4.5 years (SD, 2.8 years) had a significant reduction in lung, prostate, and colorectal cancer but not skin cancer, suggesting that selenium could not reduce the risk of skin cancer recurrence. A subsequent randomized controlled trial by the same investigators looked at the effects of selenium in men who were at high risk for prostate cancer. Participants were randomized to receive a daily oral dose of 200 mg or 400 mg of selenium or placebo. They were observed for 5 years with follow-up every 6 months. **Selenium supplementation was shown to have no effect on the incidence of prostate cancer in this population**. (Algotar AM, Stratton MS, Ahmann FR, et al., 2013)

The authors concluded that this and other studies indicate that selenium supplementation may not have a role in preventing prostate cancer.

The Selenium and Vitamin E Cancer Prevention Trial (SELECT) was the result of what appeared to be growing evidence at the time that selenium and vitamin E may reduce the risk of prostate cancer. (Helzlsouer KJ, Huang HY, Alberg AJ, et al., 2000) (Li H, Stampfer MJ, Giovannucci EL, et al., 2004)

SELECT was a large, randomized, controlled phase III trial in which researchers investigated whether daily supplementation with 200 mg of selenium, 400 IU of vitamin E (α-tocopherol), or both would lower the incidence of prostate cancer in men aged 50 years or older. (Lippman SM, Klein EA, Goodman PJ, et al., 2009)

The study began in 2001 and was terminated in 2008, 5 years earlier than originally planned. The results showed that intake of these supplements over 5.5 years did not reduce the incidence of prostate cancer or other cancers. An update reported in 2011 revealed that after an average of 7 years (5.5 years on the supplements and 1.5 years off), **there were 17 percent more cases of prostate cancer in men taking vitamin E alone than among men on placebo**. There was no increase in prostate cancer among men who were assigned to take selenium alone or vitamin E plus selenium, compared with men assigned to take placebo. (Klein EA, Thompson IM Jr, Tangen CM, et al., 2011)

In the SU.VI.MAX study, the impact of sex was observed in a randomized, placebo-controlled trial conducted in France with 7876 women aged 35–60 and 3560 men aged 45–60 years. The researchers investigated the potential cancer-preventive effects of low-dose vitamin C, β-carotene, vitamin E, selenium, and zinc. **After 7.5 years, low-dose antioxidant supplementation was associated with a reduced total cancer incidence and all-cause mortality in men but not in women**. It was thought that supplementation may have been effective only in men because of their lower baseline status of certain antioxidants, especially β-carotene. (Hercberg S, Galan P, Preziosi P, et al., 2004)

DIETARY ANTIOXIDANTS: CONFRONTING THE DILEMMA

Antioxidants, including endogenous enzymes, other molecules such as glutathione, and dietary nutrients, contribute to the body's defense systems. Dietary antioxidants also maintain the body's metabolic and structural health. Vitamin C, for example, plays a role in collagen synthesis.

Antioxidants in plant foods are complex mixtures of vitamins, minerals, and phytochemicals that may cooperate with one another in additive or synergistic ways, thereby augmenting antioxidant activity. In some studies, this may partly account for the lack of benefits found in reducing cancer risk by supplements that are purified chemicals given in doses higher than the amounts present in food.

The main essential micronutrients for antioxidant activity are lipid-soluble vitamin A, with low antioxidant activity, converted from the powerful antioxidant phytochemical β-carotene; lipid-soluble vitamin E, largely in the form of α-tocopherol; and water-soluble vitamin C (ascorbic acid). Several minerals have antioxidant activity by virtue of being part of antioxidant enzymes. These include selenium, zinc, manganese, magnesium, iron, and copper. Nonessential antioxidants in plant foods include lycopene (a carotenoid), polyphenols, and flavonoids.

Some dietary antioxidants display a double-edged sword. (Bouayed J, Bohn T., 2010)

For example, **β-carotene at low doses acts as an antioxidant with anti-inflammatory properties, but at high doses it has pro-oxidant actions** that produce proinflammatory mediators such as tumor necrosis factor-α.

In vitro studies show that vitamin C at high doses is a pro-oxidant and has a cytotoxic effect on cells by generating hydrogen peroxide. (Parrow NL, Leshin JA, Levin M., 2013)

Antioxidant supplements used in randomized trials may have had detrimental metabolic effects independent of their antioxidant actions, which may have affected study outcomes. In addition, megadoses of dietary antioxidants have also generally failed to prevent human disease, in part because they do not decrease oxidative damage as indicated by biomarkers for inflammation. (Halliwell B., 2012)

DISCUSSION

Cancer is a complex, multistage disease that entails genetic mutations and epigenetic events. Although many of the factors involved in cancer development are unknown, EMODs, including free radicals, are thought to play a role. EMODs are formed naturally during cellular metabolism.

Endogenous and dietary antioxidants were theorized to scavenge and neutralize EMODs, potentially preventing any cancer-causing effects. Supportive evidence for the role of antioxidants in cancer prevention came from experimental studies and from epidemiologic studies which showed that a high intake of fruits and vegetables rich in antioxidants reduced the risk of certain cancers. **Some more recent prospective studies have challenged these observations.** (Boffeta P, Wichmann J, Ferrari P, et al., 2010) (Griffiths K, Aggarwal BB, Singh RB, et al., 2016)

Support for dietary intake of antioxidants in lowering cancer risk originally came from observations about the protective effects of the Mediterranean diet, which is rich in antioxidant-containing fruits, vegetables, grains, and olive oil. Adherence to the diet was associated with a reduced risk of overall cancer mortality as well as a lower incidence of colorectal, breast, gastric, prostate, liver, and head and neck cancers. When the various components in this diet were assessed for their contribution, **no single ingredient or specific food category was found to mediate the beneficial effects, suggesting that they were the result of complex interactions**. (Schwingshakl L, Hoffmann G., 2016)

A systematic review by the U.S. Preventive Services Task Force of available evidence on the use of antioxidant vitamins and mineral supplements for preventing chronic diseases, including cancer, did not produce unequivocal evidence for benefit in cancer prevention. (Fortmann SP, Burda BU, Senger CA, et al., 2013)

Researchers in randomized controlled trials have examined a variety of antioxidant supplements for their impact on cancer risk or mortality. In most cases, the trials showed that isolated dietary antioxidant supplements (vitamins A, E, and C; β-carotene; and selenium), given singly or in combination, often at doses twice as high as government recommendations, did not reduce cancer risk or mortality. In some studies, β-carotene supplementation appeared to increase lung cancer risk in smokers. (Omenn GS, Goodman GE, Thorquist MD, et al., 1996)

To some, the role of antioxidants in cancer treatment remains unclear. Many patients with cancer take dietary supplements, particularly antioxidants, to reduce the toxicity associated with cancer treatment. **Despite more than two decades of research, evidence for the safety and efficacy of concurrent use of dietary antioxidants during cancer treatment remains inconclusive and dangerous.**

Some data suggest that antioxidants protect both cancer cells and healthy cells from oxidative damage. Other data suggest that antioxidants can protect normal tissue without interfering with the efficacy of chemotherapy or radiation. Several randomized trials have shown that antioxidants may help as palliative agents and may reduce the toxic side effects of cancer treatment.

The inconsistent outcomes of trials may be due to many factors that relate to study design. These include the chemical form of the antioxidant used, intervention dose, mode of administration, and study duration. In addition, baseline antioxidant status, genetic polymorphisms, gene–nutrient interactions, age, sex, type of malignancy, and health status are potential variables among study participants that are generally unaccounted for and are likely to influence results. (Bouayed J, Bohn T., 2010)

Dietary antioxidants also have the potential to interact with one another in their metabolic functions, which may limit the effectiveness of selected antioxidant supplements given in single or combined form in pharmacologic doses. (Borek C., 1997) (Niki E, Noguchi N, Tsuchihashi H, Gotoh N., 1995)

Additional well-designed, randomized, controlled clinical trials that take into account a range of potentially confounding variables are needed to further investigate the role of dietary antioxidants in cancer risk and prevention and to help better define the role that antioxidant supplements may have as adjuvants in cancer treatment.

Even following the litany of negative results listed above and in countless other scientific articles, this author (Borek) continues to be disillusioned with his conclusions. Hence, he wrote: "On the basis of current available information, intake of antioxidant-rich fruits and vegetables is encouraged for health maintenance and for its possible role in lowering disease risk."

RMH Note: For some reason unbeknownst to me, many so-called scientists cannot bring themselves to face the reality that the free radical theory has been debunked by decades of reliable scientific data.

There appears to be a continued public misconception (encouraged by the supplement industry) that free radicals are bad, and that antioxidants are good.

Until recently, the thinking had been that the more antioxidants, the less oxidative stress, because all of those lonely electrons would quickly get paired up before they had the chance to start mucking things up in our cells. But that thinking has changed.

Drs. Cleva Villanueva and Robert Kross published a 2012 review titled "Antioxidant-Induced Stress" in the International Journal of Molecular Sciences.

A person taking high dose antioxidants might actually be doing the exact opposite of what he or she is trying to accomplish with the supplement? Yes, they said. But it gets even worse.

Free radicals are a natural byproduct of aerobic metabolism in the cells (energy production using oxygen), which ramps up during physical activity. The free radicals produced by this increase in metabolism signal the cell to make its own home-made antioxidants. These endogenous antioxidants are very important and may be responsible for many of the health benefits associated with physical activity.

A 2014 review published in Nutrition and Food Science called for more research on the subject, but ultimately concluded that high-dose antioxidant supplements can effectively "abolish the beneficial effects of exercise". Researchers think that high levels of a single antioxidant (like the 1,000 mg of vitamin C in a packet of Emergen-C) can snatch up all the free radicals produced by exercise before they have a chance to trigger the synthesis of those beneficial endogenous antioxidants.

Barbara Demmig-Adams, professor of ecology and evolutionary biology at the University of Colorado in Boulder, and one of the authors of the 2014 paper wrote, "I think it's a really important realization that the much-maligned radicals have a job to do in our bodies and that single high-dose supplements can do more harm than good. . . Our review on antioxidants and exercise is just the tip of the iceberg. **There is a** real paradigm shift . . . in the biomedical research area that is causing pioneers to ask, 'could 50 years of research be wrong?"

Scientists are never 100 percent sure about anything. But marketers seem to be so certain about everything they tell us.

A History of Mistakes

The following was modified or excerpted from: (Villanueva and Kross, 2012)

Antioxidant-Induced Stress

Antioxidants are among the most popular health-protecting products, sold worldwide without prescription. Indeed, there are many reports showing the benefits of antioxidants but only a few question the possible harmful effects of these "drugs". The normal balance between antioxidants and free radicals in the body is offset when either of these forces prevails. The available evidence on the harmful effects of antioxidants is analyzed in this review. In summary, a hypothesis is presented that "antioxidant-induced stress" results when antioxidants overwhelm the body's free radicals.

1. Introduction

The emergence of aerobic organisms was a consequence of oxygen becoming a significant component of the earth's atmosphere. The evolution of aerobic organisms was probably the result of their gradual adaptation to oxygen. Antioxidant enzymes probably facilitated part of such adaptation, since strict anaerobes have neither superoxide dismutase (SOD) nor catalase, whereas aerotolerant anaerobes have small levels of those enzymes. (McCord J.M., Keele B.B., Jr, Fridovich I., 1971)

Aerobic organisms use oxygen to produce the chemical energy they need to survive. Metabolic pathways of aerobic organisms lead to the obligate production of reactive oxygen species (ROS) and reactive nitrogen species (RNS), which are not necessarily bad. Indeed, **reactive species are actually needed as signal transduction elements in processes that are essential for life, such as cell growth, differentiation, preconditioning and cell death**. (Laranjinha J., 2009) (Gutierrez J., Ballinger S.W., Darley-Usmar V.M., Landar A., 2006)

They also participate in functions such as phagocytosis and stress response [4]. This echoes Barry Halliwell's **"All aspects of aerobic life involve free radicals and antioxidants—you cannot escape them, nor should you wish to."** (Halliwell B., 2009)

Free radicals are molecules or atoms characterized by possessing an unpaired electron that is always seeking a counterpart to reside inside the parent home or in the home of the other electron. Free radicals can be produced during cellular processes or as a result of the interaction of free radicals, reactive species, or free radical-generating systems with other molecules.

Many of the EMODs and RNS are free radicals. There are also macromolecular radicals that are formed by the effect of oxygen, drug-induced free radicals, irradiation or free radicals on such macromolecules as protein and lipids. For example, it has been demonstrated that γ-irradiated proteins produce electron paramagnetic resonance (EPR)—detectable superoxide and hydroxyl radicals, in the presence of ferrous iron, via their intermediate conversion to hydroperoxides.

According to Halliwell and Gutteridge an antioxidant is "any substance that, when present at low concentrations compared to those of an oxidizable substrate, significantly delays or prevents oxidation of that substance." (Halliwell B., Gutteridge J.M., 1995)

Just as reactive species are not necessarily bad, antioxidants are not necessarily good.

Antioxidants are a kind of matchmaker midwife trying to give or receive an electron to complete the other's "pair", **but by making a proper pair, antioxidants can themselves become incomplete, and behave as free radicals**; i.e., reactive and looking to pair with an unpaired electron.

It is known that those "reactive" antioxidants can be stabilized (recycled) by other antioxidants, which in turn become "reactive" and can be recycled in a cascade of "reactivity mitigation". It is all a matter of equilibrium. It well is accepted that oxidative stress is produced when the equilibrium between reactive species and antioxidants is tilted in favor of the former. However, the equilibrium can be broken also if antioxidant levels exceed those of the reactive species.

Dündar proposed the term "antioxidative stress" for such disequilibrium. (Dündar Y., Aslan R., 2000)

2. Physiological Effects of Reactive Species

Reactive species participate in different functions and play a role as signal transduction elements in many physiological events. ROS and RNS participate in signal transduction of cytokine receptors, tyrosine receptor, serine/threonine kinases, G protein-coupled receptors, ion-channel linked receptors in response to angiotensin II, cytokines, glutamate, epidermal growth factor, vascular endothelial growth factor, tumor necrosis factor α and platelet derived growth factor. (Laranjinha J., 2009)

Free radicals and EMODs participate in the redox regulation of activator protein 1, nuclear factor κ B (NFκ-B), cyclic response element-binding protein (CREB), nuclear factor E2-related factor 2 (Nrf2) and p53 ("guardian of the genome"). Superoxide and hydrogen peroxide activate extracellular signal-related protein kinase (ERK), protein kinase B, mitogen-activated protein kinases (MAPKs) and insulin receptor kinase [15,21–23]. (Valko M., Leibfritz D., Moncol J., Cronin M.T., Mazur M., Telser J., 2007)

The effects of reactive species are dose-dependent. Incubation of Jurkat T-cells with **increasing concentrations of H2O2 produced proliferation (0.7 µM), apoptosis (1.0–1.3 µM) or necrosis (>3 µM)**. (Cadenas E., 2004)

EMODs participate in proteolysis into the proteasome. **H2O2 increases proteolysis at concentrations of 20–400 µM, whereas it inhibits proteolysis at higher, mM concentrations, producing an accumulation of oxidized proteins.**

Preconditioning is one of the areas where the dose-dependent physiological effects of reactive species are evident. Repeated or unique application of injurious stimuli, at intensities below the threshold of damage, activates endogenous mechanisms that afford protection when subsequent major injurious stimuli are applied. **This phenomenon is known as preconditioning.** (Gu G.J., et al., 2008)

Hyperbaric oxygen, short ischemia, small doses of lipopolysaccharides, restraint stress, hypoxia, hyperthermia and moderate aerobic exercise are some of the stimuli producing preconditioning.

An excess of ROS are produced during acute or intense exercise. (Syu G.D., Chen H.I., Jen C.J., 2011) (Fisher-Wellman K., Bell H.K., Bloomer R.J., 2009

It was thought that mitochondria were the primary site of superoxide production during exercise, because oxygen consumption in respiration is increased. However, it is now known that superoxide is produced mainly through NADPHox, XO and phospholipase A2 (PLA2) during intense exercise. (Gomez-Cabrera M.C., Domenech E., Vina J., 2008)

Chronic moderate exercise suppresses oxidative stress produced by acute intense exercise. The mechanism of such adaptation seems to involve, as has been demonstrated in different studies, activation of NFκ-B by ROS and the consequent expression of nitric oxide synthases and antioxidant enzymes. Indeed, **the incubation of cells with hydrogen peroxide induces the expression of antioxidant enzymes**. (Veal E.A., Day A.M., Morgan B.A., 2007)

3. Pro-Oxidant Effects of Antioxidants

As mentioned above, antioxidants become "unstable" and "reactive" when they lose or receive electrons in the presence of reactive species. **In these conditions antioxidants exhibit pro-oxidant effects** and can become harmful. Their redox potential (tendency to acquire electrons and thereby be reduced) could be related to the toxic effects of antioxidants. The redox potentials of some antioxidants of interest are: 0.65 V for β-carotene, 0.50 V for Vitamin E, 0.25–0.50 V for flavonoids, 0.25 V for uric acid and 0.01 V for Vitamin C.

It has been said that a network of antioxidants is necessary. Antioxidants should be interconnected in such a way that they can turn off each other's reactivity after interacting with reactive species. Precisely; **Damiani defined a good antioxidant as one that produces a low oxidant reactivity with a low capacity to produce peroxidation**. (Damiani E., Astolfi P., Carloni P., Stipa P., Greci L., 2008)

α-tocopherol (α-TC, Vitamin E) produces α-tocopheroxyl radical (α-TC.) when it reacts with reactive species such as peroxynitrite or superoxide. α-TC. is then recycled to α-TC by other antioxidants such as ascorbic acid (Vitamin C) and glutathione.

As soon as ascorbic acid recycles Vitamin E, it is transformed to the ascorbyl radical, which has a lower reactivity than α-TC. α-TC. is also recycled to α-TC by β-carotene. **Therefore, it is important that Vitamin E be administered with another antioxidant.**

It is also important to have significant levels of endogenous antioxidants to allow for the recycling of "reactive" antioxidants. Some conditions reduce endogenous antioxidants and could potentially affect the recycling of Vitamin E (or another antioxidant), leaving α-TC. available to produce lipoperoxidation. One such condition is the reduction of ascorbic acid in smokers.

Ascorbic acid also recycles oxidized glutathione (GSSG) to reduced glutathione (GSH), and carotenoid radicals to carotenoids. Dihydro-lipoic acid (DHLA), the reduced form of lipoic acid (an antioxidant) recycles GSH from GSSG and the ascorbic ascorbyl radical. (Moini H., Packer L., Saris N.E., 2002)

The pro-oxidant or antioxidant effects of some antioxidants depend on their concentration. For example, lipoic acid and DHLA inhibit nitro-tyrosine formation by peroxynitrite at a 0.01–0.05 mM concentration, whereas promote mitochondrial permeability transition at 0.1 mM, increase of glucose intake at 0.25 M, and stimulate calcium mitochondria release at 1 mM. β-carotene protects DNA oxidative damage at 1–3 μM, **whereas it oxidizes DNA at 4–10 μM. High concentrations of β-carotene inhibit the growth of cancer cells in culture and promote their apoptosis.**

Some antioxidants exhibit pro-oxidant activities in the presence of transition metals. Such is the case of ascorbic acid that is transformed into an ascorbyl radical and hydroxycinnamic acid whose phenolic groups are transformed into phenoxy reactive groups. (Poljsak B., et al., 2005) (Maurya D.K., Devasagayam T.P., 2010)

Retinal and carotenoid radical are produced by the oxidative cleavage of β-carotene in the presence of oxygen and ferrous ion.

Oxidative cleavage of β-carotene is also produced by many other oxidative stimuli: ultraviolet light (artificial and sun light), heat, free radicals, and ozone. Carotenoid breakdown products

(CBP) were produced in ferrets by cigarette smoke exposure. The effect was prevented by α-TC and ascorbic acid.

Reactive species produced by the interaction of an antioxidant with a free radical such as ROS or RNS exhibit potentially toxic pro-oxidant effects. CBP inhibit mitochondrial respiration in the rat lung, brain and liver, stimulate apoptosis in neutrophils, oxidize DNA [60] and up-regulate advanced glycation end products.

In summary, **the pro-oxidant and potential harmful effects of an antioxidant depend on (a) its concentration; (b) its redox potential; (c) the presence of other antioxidants; (d) the presence of transition metals, (e) the activity and concentrations of endogenous antioxidants**. And, even though not yet demonstrated, the genetic background could also contribute to the harmful effects of antioxidants.

4. Clinical Trials with Antioxidants

The validity of antioxidant therapy has been questioned, having been pointed out that even though several diseases have been related to oxidative stress, antioxidant therapy does not change the natural course of diseases. (Halliwell B., 2009) (Halliwell B., 2008) (Long L.H., Hoi A., Halliwell B., 2010) (Halliwell B., 2007) (Halliwell B., Lee C.Y., 2010) (Gutteridge J.M., Halliwell B., 2010)

Some authors have commented that antioxidant therapy has only been validated in experimental models of disease and cell culture.

However, different factors cause artifacts in cell culture (e.g., high oxygen levels compared to the pO2 of tissues in the organism, components of the cell culture media, high concentrations of antioxidants).

In the clinical arena one of the problems is the biomarkers to evaluate oxidative stress. (Halliwell B., Lee C.Y., 2010)

It has been postulated that a combination of low doses of antioxidants could only be useful in patients suffering from antioxidant deficiencies. (Halliwell B., Lee C.Y., 2010)

It remains unknown if patients with gene polymorphisms affecting the structure and function of antioxidant enzymes could have different responses to antioxidants.

What is interesting is that when no effects or negative results of antioxidants are reported, there is often an alternate conclusion that antioxidants can be dangerous. For example: the method used to evaluate oxidative stress, treatment strategy (preventive strategy, doses, combination of antioxidants, duration), the stage of the disease, the follow up period, and number of patients. Interestingly, in a recent review, **Ristow and Schmeisser suggest that life span could be extended by oxidative stress.** (Ristow M., Schmeisser S., 201)

Also, please remember that Chang showed that antioxidants can accelerate the aging process.

A summary of clinical trials with antioxidants, as well as possible explanations are shown below.

4.1. **Beneficial Effects of Antioxidants**

Nurses' Health Study (NHS). This was a cohort study of more than 87,000 U.S. nurses, ages 34–59, with no history of cardiovascular disease (CVD). Follow up period: 8 years. Results: Supplemental Vitamin E (at least 100 IU/day, for at least 2 years) was associated with reductions of 40% or more in the risk of coronary heart disease (CHD). (Bowman T.S., Bassuk S.S., Gaziano M., 2007) (Stampfer M.J., Hennekens C.H., Manson J.E., et al., 1003)

Health Professional Follow up Study (HPFS). This was an observational study of around 40,000 U.S. male health professionals, ages 50–75, with no CHD, diabetes or hypercholesterolemia. Supplemental Vitamin E (at least 100 IU/day, for at least 2 years) was associated with a significant reduction of CHD risk. (Bowman T.S., Bassuk S.S., Gaziano M., 2007) (Rimm E.B., Stampfer M.J., Ascherio A., et al., 1993)

The "Established Populations for Epidemiologic Studies of the Elderly" program. This involved 11,178 U.S. men and women, ages 67–105. Results: There is a significant decreased risk of CHD mortality among those taking vitamin supplements. (Bowman T.S., Bassuk S.S., Gaziano M., 2007) (Losonczy K.G., Harris T.B., Havlik R.J., 1996)

The first National Health and Nutrition Examination Survey (NHANES I). This involved 11,348 U.S. women and men, ages 25–74. Conclusion: Vitamin C intake was inversely associated with all causes of mortality and CVD in men, but not in women. (Hasnain B.I., Mooradian A.D., 2004)

Scottish Heart Health study. This involved 3833 women and 4036 men, ages 40–59, with no heart disease at the beginning of the study. Vitamin C and β-carotene intake were associated with reduced CHD, only in men. (Hasnain B.I., Mooradian A.D., 2004) (Todd S., Woodward M., Tunstall-Pedoe H., et al., 1999)

In a recent meta-analysis Myung et al. evaluated 22 case-control reports (from 274 articles), including 10,073 patients with a cervical neoplasm risk. They found preventive effects of Vitamins B12, C, E and β-carotene on cervical neoplasms. (Myung S.K., Ju W., Kim S.C., 2011)

Alpha tocopherol, Beta carotene cancer prevention study (ATBC). This involved 27,111 Finnish male smokers, ages 50–69. Follow up: 16–19.4 years. Results: **A higher αTC intake was associated with a lower pancreatic and prostate cancer risk. In the same study high flavonoid intake was also related to lower pancreatic cancer risk.** (Stolzenberg-Solomon R.Z., Sheffler-Collins S., Weinstein S., et al., 2009) (Weinstein S.J., Wright M.E., Lawson K.A., et al. 2007) (Bobe G., Weinstein S.J., Albanes D., et al., 2008)

4.2. Antioxidants Do Not Change the Evolution of Diseases Related to Oxidative Stress

20,536 patients with hypertension at high cardiovascular risk were followed for five years. **Treatment with ascorbic acid, Vitamin E and β-carotene neither reduced blood pressure nor changed mortality or morbidity.** (Pechanova O., Simko F., 2009) (Heart Protection Study Collaborative Group. Mrc/bhf, 2002)

Rotterdam study. This involved 4,802 Dutch men and women, ages 55–95, with no history of myocardial infarction (MI). Follow up: **After 4 years, no association was observed between Vitamin E intake and the risk of MI.** (Hasnain B.I., Mooradian A.D., 2004) (Klipstein-Grobusch K., Geleijnse J.M., den Breeijen J.H., et al., 1999)

Scottish Heart Health Study. No protection was associated with Vitamin E. Vitamins C, E and β-carotene had no effect on all-cause mortality. (Hasnain B.I., Mooradian A.D., 2004)

Heart Protection Study. This included 20,536 men and women with CHD or diabetes, ages 40–80. Follow up: **After 5 years, the intake of Vitamin E (600 mg/day), Vitamin C (250 mg/day) and β-carotene (20 mg/day) had no effect**. (Hasnain B.I., Mooradian A.D., 2004) (

Primary Prevention Project. This program involved 4,495 men and women, with an average age of 64 and at least one risk factor for CHD, who took a low dose of aspirin (100 mg/day) and Vitamin E (300 mg/day). **The study was stopped early because other trials had demonstrated protection with aspirin. Vitamin E had no effect**. (Hasnain B.I., Mooradian A.D., 2004)

Heart Outcomes Prevention Evaluation Study (HOPES). This included 2545 women and 6999 men, ≥55, with CHD or diabetes and one risk factor for atherosclerosis. They received Vitamin E (400 IU/day) or a placebo, angiotensin-converting enzyme inhibitor or a placebo. A follow up after 4.5 years indicated that Vitamin E had no effect. (Hasnain B.I., Mooradian A.D., 2004) (Yusuf S., Dagenais G., Pogue J., et al., 2000)

Gruppo Italiano per lo Studio della Sopravvivenza nell'Infarto Miocardico (GISSI). This included 11,324 Italian men and women who had survived a myocardial infarction within the 3 previous months. **They received Omega-3 oils (1 g/day) and/or Vitamin E or no treatment. A follow up after 3–5 years showed that Vitamin E had no effect**. (Yusuf S., Dagenais G., Pogue J., et al., 2000)

Rytter et al. studied 40 patients with type 2 diabetes, who were randomly assigned to one of two groups, one treated with antioxidants extracted from fruits, berries and vegetables (tocopherols, carotenoids and ascorbate), and the other treated with placebo. **The authors found that 12 weeks of treatment did not change the metabolic profile, neither oxidative nor inflammatory biomarkers, even though antioxidant systemic levels had increased**. (Rytter E., Vessby B., Asgard R., et al., 2010)

Recently, Suksomboon et al. analyzed 9 trials including 418 type 2 diabetic patients who were treated with Vitamin E for 8 weeks. They found that **Vitamin E did not improve glycemic control unless the patients began their treatment under bad metabolic control and with a Vitamin E deficiency**. (Suksomboon N., Poolsup N., Sinprasert S., 2011)

Arain et al. analyzed four clinical trials to evaluate the effects of Vitamin E on colorectal cancer. The trials included 94,069 patients (47,029 received Vitamin E) and had a follow up period of 4 years. The authors concluded that Vitamin E had no effect on colorectal cancer. (Arain M.A., Abdul Qadeer A., 2010)

4.3. Harmful Effects of Antioxidants

We summarize here some of the published studies that show the harmful effects of antioxidant supplements in humans.

In a study made in Australia, 69 hypertensive patients with an ambulatory systolic pressure of >125 mmHg received treatment with Vitamin C (500 mg/day) and grape seed polyphenols (1000 mg/day) for 6 weeks. **At the end of the treatment their systolic and diastolic pressures increased and there were no changes in their endothelium dependent vasodilation or oxidative stress biomarkers**. (Ward N.C., Hodgson J.M., Croft K.D., 2005)

HDL Atherosclerosis Treatment Study (HATS). 160 men (<70) and women (<63), with low HDL and triglycerides <400 mg/dL. Patients were assigned to one of 4 groups: Simvastatin + niacin with or without Vitamin E (800 IU/day), Vitamin C (1000 mg/day), β carotene (25 mg/day) and selenium (100 µg/day); Antioxidants or placebo. **A follow up after 3 years showed that antioxidant treatment blunted HDL elevation produced by simvastatin + niacin.** (Hasnain B.I., Mooradian A.D., 2004) (Cheung M.C., Zhao X.Q., Chait A., 2001)

The Prostate, Lung, Colorectal and Ovarian cancer Screening Trial (PLCO) was a prospective study of 25,400 postmenopausal U.S. women, ages 55–74, who were followed for 10 years. **The results showed that the risk of breast cancer increased significantly (by 20%) in women who had a folic acid supplement (≥400 µg/day) whereas food folate intake was not associated with an increased risk.** (Kim Y.I., 2006)

Beta carotene and Retinal Efficacy Trial (CARET). This involved 18,314 U.S. women and men on β carotene (30 mg/day) and Vitamin A (25,000 IU/day) or a placebo. **The study had to be stopped two years prematurely because Vitamin-treatment was associated with a 28% greater incidence of lung cancer and 17% more deaths than placebo treatment.** (Omenn G.S., Goodman G.E., Thornquist M.D., 1996) (Hasnain B.I., Mooradian A.D., 2004) (Hasnain B.I., Mooradian A.D. Recent trials of antioxidant therapy: What should we be telling our patients? Cleve. Clin. J. Med. 2004, 71:327–334)

Alpha Tocopherol Beta Carotene Cancer Prevention Study (ATBC). This covered 29,133 male smokers, ages 50–69, taking α-TC (50 mg/day) and/or β-carotene (30 mg/day). **There was an 8% greater mortality in those men taking the supplements than the men treated with the placebo.** (Hasnain B.I., Mooradian A.D., 2004) (Albanes D., Heinonen O.P., Taylor P.R., 1996) (The Alpha-Tocopherol Beta Carotene Cancer Prevention Study Group, 1994)

Bjelakovic et al. published a meta-analysis of antioxidant supplements for the prevention of gastrointestinal cancer. They selected 7 high quality randomized trials (of the 14 trials examined), including 131,727 patients. **They found that antioxidants significantly increased overall mortality and did not prevent gastrointestinal cancer.** (Bjelakovic G., Nikolova D., Simonetti R.G., et al., 2004)

Myung et al. analyzed 31 of 3,327 articles searched, including 22 trials with 161,045 patients (88,610 treated with antioxidants). **<u>The authors concluded that antioxidants had no preventive effects on cancer. However, when they evaluated a subgroup of four controlled trials they found that patients receiving antioxidants had a significant increase in bladder cancer</u>**. (Myung S.K., Kim Y., Ju W., Choi H.J., 2010)

Antioxidants Vitamin C (12.5 mg/Kg) and N-acetylcysteine (10 mg/Kg) significantly increased the oxidative stress produced by acute exercise in healthy subjects. The effect was attributed to the conversion of ascorbic acid into the ascorbyl radical by reactive species generated during exercise.

Ristow et al. studied insulin sensitivity, transcriptional regulators of insulin sensitivity and gene expression of SODs, GPx and catalase in 19 trained and 20 untrained healthy young men who were treated with Vitamin E (400 IU/day) + Vitamin C (1000 mg/day) or placebo and were trained for four weeks. They found that exercise increased insulin sensitivity as well as the gene expression of insulin sensitivity transcriptional regulators and antioxidant enzymes. **The effects**

of exercise were blocked by antioxidant supplements. (Ristow M., Zarse K., Oberbach A., et al., 2009)

Recently, Peternelj and Combes evaluated 23 studies that **reported negative effects of antioxidant supplements on the beneficial effects produced by chronic exercise. They found that antioxidants interfere with vasodilation and increased insulin signaling produced by exercise**. (Peternelj T.T., Coombes J.S., 2011)

Recently, the Iowa Women's Health Study results were published. The study included 38,772 women, who were older than 60 in 1986. Vitamin and mineral supplement intake was self-reported at different periods. <u>**The authors concluded that dietary Vitamin and mineral supplements may be associated with increased mortality**</u>, with the effect being worse when the supplements were accompanied by an iron supplement. (Mursu J., Robien K., Harnack L.J., et al., 2011) (Mursu J., Robien K., Harnack L.J., Park K., Jacobs D.R., Jr Dietary supplements and mortality rate in older women: The Iowa women's health study. Arch. Intern. Med. 2011; 171:1625–1633)

It is known that smoking produces oxidative stress and cancer; and also known, as mentioned above, **that an antioxidant is converted to a reactive species when it interacts with a free radical. A second antioxidant, by reacting with the "activated" antioxidant, can mitigate its free radical effects and recycle it as an antioxidant**. These could explain the results of the ATBC study. Vitamin E alone did not produce any change, but the incidence of lung cancer and mortality increased in those participants who received Vitamin E and β-carotene. <u>**Moreover, the CARET study, that looked for effects of Vitamin A and β-carotene in women and men who had worked with asbestos or were smokers (both conditions are related to cancer), had to be stopped prematurely because of the clear correlation of increased lung cancer incidence and mortality**</u>. (Omenn G.S., Goodman G.E., Thornquist M.D., Balmes J., et al., 1996)

Another mechanism that seems to affect the final result of antioxidant supplementation is the normal responses of the organism to oxidative stress (OS). In the first part of this review, it was mentioned that acute exercise produces OS.

Chronic exercise reduces the OS produced during acute exercise by activating NFκ-B, which leads to gene expression of antioxidant enzymes, and NOSs. The phenomenon is termed "preconditioning". It has been postulated that antioxidant supplements, by blocking free radicals and therefore the signal-transduction pathways that they activate during chronic exercise, inhibit such preconditioning.

5. Conclusions

Even though antioxidants have been considered beneficial, <u>**a significant number of clinical trials have shown harmful effects. The most extreme results evidenced increased mortality in individuals taking antioxidant supplements**</u>.

Although the mechanisms are not explained in those trials, the results suggest that disrupting the delicate balance between antioxidants and reactive species produces antioxidant-induced stress, if the antioxidants overwhelm the physiological production of reactive species.

RMH Note: I believe that the most important observation is that excess antioxidants produce an EMOD insufficiency.

Though these conclusions remain speculative, the best advice would be to ingest antioxidants from food sources rather than from self-prescribed supplements. The most probable way of increasing the endogenous antioxidant defense would be by practicing moderate aerobic exercise, as an everyday healthy routine. Gene polymorphisms, which condition endogenous antioxidant deficiencies, would be the exception to such advice.

Antioxidant Supplements Not Beneficial in Cancer

Overall, many randomized controlled trials (RCTs) have not provided evidence that dietary antioxidant supplements are beneficial in primary cancer prevention.

In addition, **a systematic review of the available evidence regarding the use of vitamin and mineral supplements for the prevention of chronic diseases, including cancer, conducted for the United states Preventive Services Task Force (USPSTF) likewise found no clear evidence of benefit in preventing cancer.** (Fortmann SP, Burda BU, Senger CA, et al, 2013) (Fortmann SP, Burda BU, Senger CA, et al, and mineral supplements in the primary prevention of cardiovascular disease and cancer: an updated systematic evidence review for the U.S. Preventive Services Task Force. Annals of Internal Medicine 2013)

Several RCTs found that people who took antioxidant supplements during cancer therapy had worse outcomes, especially if they were smokers. (Lawenda et al, 2007) (Lawenda BD, Kelly KM, Ladas EJ, et al, 2008)

Bardia et al, showed that regular intake of supplements containing beta-carotene may increase cancer incidence and cancer deaths among smokers, according to a large systematic review and meta-analysis published in 2008. The same report also showed that selenium supplementation may have cancer-fighting effects in men, while vitamin E supplementation had no effect on cancer incidence and mortality (Bardia et al, 2008).

High-dose vitamin C supplements diminish the benefits of exercise in athletic training and disease prevention.

(Adams, R B; Egbo, K Nkechiyere; Demmig-Adams, Barbara., 2014)

The purpose of this review is to summarize new research indicating that high-dose supplements of the antioxidant vitamin C can interfere with the benefits of physical exercise for athletic performance and the risk for chronic disease. Design/methodology/approach - This article reviews current original literature on the regulation of human metabolism by oxidants and antioxidants and evaluates the role of exercise and high-dose vitamin C in this context. The presentation in this article aims to be informative and accessible to both experts and non-experts. Findings - **The evidence reviewed here indicates that single, high-dose supplements of the antioxidant vitamin C abolish the beneficial effects of athletic training on muscle recovery and strength as well as abolishing the benefits of exercise in lowering the risk for chronic disease.** In contrast, an antioxidant-rich diet based on regular foods apparently enhances the benefits of exercise.

RMH Note: This helps explain the fact that exercise, which produces high levels of EMODs, decreases the risk of cancer and it does so because of sufficient levels of EMODs to kill cancers. Antioxidants will negate the positive effects of exercise.

RMH Note: This confirms that antioxidants are useless in preventing cardiovascular diseases and lends credibility to the increase in heart attacks seen in people who take excess amounts of antioxidant supplements.

CANCER

Antioxidants May Make Cancer Worse - Scientific American
https://**www.scientificamerican.com**/article/**antioxidants**-may-make...
Antioxidants May Make Cancer Worse. ... of the **antioxidant** beta-carotene increased the risk of lung ... treated mice caused levels of RhoA to **increase** in their ...

**Can you recall our major breakthroughs
in our war against cancer?
... neither can I!
Mankind must adopt
prooxidant cancer therapies
for truly innovative advances.**
R.M. Howes, M.D., Ph.D.
3-6-16

Cancer patients should avoid the routine use of antioxidant supplements during radiation and chemotherapy because the supplements may reduce the anticancer benefits of therapy, researchers concluded in a commentary published online May 27, 2008 in the Journal of the National Cancer Institute. The Royal College of Radiologists advise patients to avoid antioxidant supplements, especially in high doses, during their conventional cancer therapy.

Oversupplementation may actually produce an environment that is beneficial to the tumor and allow it to survive.

Conversely, **even modest quenching or reduction of EMODs by dietary antioxidants could block initiation or completion of apoptosis.**

"Can vitamin and herbal supplements reduce the adverse effects of cancer treatment, decrease the risk of cancer recurrence or improve a patient's chances of survival? **We don't really know.** Research into these matters has been minimal," said senior author Cornelia (Neli) Ulrich, Ph.D., an associate member of the Hutchinson Center's Public Health Sciences Division. "While supplement use may be beneficial for some patients, such as those who cannot eat a balanced diet, **research suggests that certain supplements may actually interfere with treatment or even accelerate cancer growth**," she said in an article published Feb. 1, 2008 in the *Journal of Clinical Oncology.*

In December of 1999, Salganik pointed out that, "**Almost all anticancer drugs kill cancer cells by way of apoptosis, and antioxidants like vitamin A and vitamin E dramatically reduce apoptosis in cancer cells." Patients should therefore avoid taking any more than a normal amount of these vitamins during chemotherapy treatment.**

Harman's free radical theory

**The free radi-crap theory is
the dropped end product of
Kerplunk Sphincteric Productions.**
R. M. Howes, M.D., Ph.D.
12-12-09

EMODs and Spontaneous Cancer Regression

EMODs Cause Spontaneous Regression of Cancer (Melanoma)

My first clue about our ability to naturally kill cancer was related to "spontaneous regression or remission," which is defined as cancers that shrink or disappear completely, without any conventional or alternative anticancer therapy. It is well-established in the medical literature (hundreds of descriptions) and amongst oncologists.

The most frequently cited cancers that may experience a spontaneous remission include kidney and testicular cancers as well as lymphoma and melanoma (estimated at 1 out of every 400 cases). Perhaps the best studied group is a type of lymphoma referred to as "low-grade, B-cell, Non-Hodgkin's lymphoma." It is well established that approximately 20% of patients diagnosed with this type of lymphoma will experience a spontaneous shrinkage of their disease. It is for this reason that oncologists do not treat these types of lymphoma unless they are causing bothersome symptoms for the patient.

In the *New York Times* (Oct 27, 2009), Gina Kolata wrote an article entitled, "Cancers Can Vanish Without Treatment, but How?" The impetus for her article was recent medical publications on screening tests for breast cancer and prostate cancer (mammography and PSA testing). Kolata wrote, "Besides finding tumors that would be lethal if left untreated, screening appears to be finding many small tumors that would not be a problem if they were left alone, undiscovered by screening. They were destined to stop growing on their own or shrink, or even, at least in the case of some breast cancers, disappear."

I believe that EMODs are, at least to a significant degree, the basis of this phenomenon. They are primarily responsible for EMOD-induced apoptosis (cell suicide), along with a vast complexity of biochemical entities.

Please remember that EMODs can be blocked by antioxidants, which can interact when combined in what is called a "cocktail effect." Others refer to it as a "synergism." Actually, my work shows that combining the antioxidant vitamins increases their harmful consequences, which I presented and discussed in *Death In Small Doses?*

The developed nations have the highest rates of common cancers, such as prostate, breast and lung and it appears that we are transferring this to the lesser developed nations. **Antioxidants are in hundreds of food products and manufactured goods in the industrialized nations.** Are they causal of these diseases? Are they related to the BPAs (bisphenols) used in the plastics industry and which is an antioxidant used to prevent polymerization? We now have excellent leads to answers concerning antioxidants.

Are developed nations like the wealthy Romans, who could afford lead-laden plates and pots and subsequently died of lead poisoning, because they could afford it? The poor were spared and this may be analogous to under developed nations not being able to afford the harmful products (antioxidants) of the industrialized nations (contaminated baby bottles, drink cans, water lines, fortified foods, antioxidant vitamin supplements, etc.).

There are those who say that "cancer is a pathway disease" and I believe that EMODs will lead the way to the cure and prevention of cancer. Redox (oxidation/reduction) pathways will lead us to the door of discovery and I am trying to open and go through that door.

Chemotherapies that Produce Free Radicals, EMODs (Partial List)

Anthracyclines Such as:
Doxorubicin (Adriamycin®, Doxil®)
Daunorubicin (Cerubidine®)
Epirubicin (Ellence®)
Mitomycin (Mutamycin®)
Bleomycin (Bleoxane®)
Podophyllum agents such as:
Etoposide (VP-I6, Vespid®)
Teniposide (Vumon®)

So, we will start by focusing on my fundamental and powerful prooxidant approach to handling cancer successfully.

Prooxidant Cancer Therapies

Prooxidants, such as oxygen free radicals or reactive oxygen species, have arguably been blamed for cancer causation and some unscrupulous marketers have brought discredit to oxygen-based therapies and disregard for oxidative centered treatments.

In contrast, a worldwide review of currently effective tumoricidal methods reveals a "commonality of cures," in that many successful cytotoxic agents, procedures or methods have been shown to involve and proceed in a significant way via prooxidants.

Points of convergence occur with chemotherapy, radiation therapy, megadose intravenous vitamin C therapy, photodynamic therapy, sonodynamic therapy, the Howes' singlet oxygen tumoricidal system, ozone therapy, hyperbaric oxygen therapy and hydrogen peroxide therapy.

Contrary to the teachings of the invalidated free radical theory, certain prooxidants may offer unique tumoricidal properties and approach the Holy Grail (**HO$_2$LY GRΔIL**) for cancer treatments, allowing for selective killing of cancer cells while sparing normal cells.

RMH Note: I now know that cancer can be prevented and/or reversed by increasing the systemic oxidative capacity to a point I call "oxidative bliss," which triggers cancer cell death.

The common biochemical thread is wrapped around prooxidation and EMOD induced apoptosis, which is in essence states that "electronically modified oxygen derivatives (EMODs, formerly known as ROS, reactive oxygen species), when in proper concentrations and in the proper cellular locations, will cause the cancer cell to commit apoptotic suicide, thus ending its devastating course of growth and dissemination."

Arguments abound as to the genetic or environmental factors causing cancer but **there is very little argument against the effectiveness of EMOD induced apoptosis.**

EMODs Serve as Pro-Survivor Agents for the Victims of Cancer and as Pro-Death Agents for Cancerous Cells.

Apoptosis or cellular suicide is one of the most important means of eliminating precancerous and cancerous cells from the body. The two-other means are necrosis and autophagy.

In many of my books, and strictly based on the scientific literature, I have emphasized the fact that electronically modified oxygen derivatives (EMODs, formerly called reactive oxygen species, ROS or oxygen free radicals) play a crucial role in the killing (apoptosis) of cancerous cells. I refer to this as a "radical way to stop cancer."

The primary means that the human body uses to rid itself of precancerous and cancerous cells is mediated by electronically modified oxygen derivative (EMOD) induced apoptosis. This crucial process can be blocked or negated by either small molecular antioxidants or by antioxidant enzymes. Cellular proliferation, cellular arrest and cellular suicide appear to be modulated by relative concentrations of EMODs.

In spite of claims to the contrary, I have also stressed and highlighted the point that so-called "antioxidants" can effectively shield or protect cancer cells from being killed by exposure to EMOD prooxidants.

I believe that curing and preventing cancer revolves around the two-following essential, evidence based, scientific conclusions: EMODs kill cancer cells and antioxidants protect cancer cells. With this knowledge, the time has come to eradicate cancer from our midst and to protect those who have not yet fallen victim to the current global cancer pandemic.

In my books, I present many *in vitro* cases whereby EMODs are proven cancer cell killers. Naturally, one may ask, "Does this concept hold true for *in vivo*, human cancer cells?" The answer is a resounding, "Yes."

My chapters on chemotherapy, radiation therapy, photodynamic therapy, IV megadoses of vitamin C and the Howes Singlet Oxygen tumoricidal system, exercise and the naked mole rat all involve EMOD induced cancer cell apoptosis. Thus, **there are millions of *in vivo* human examples of cancer cell killing and suppression to support the cell culture experiments**.

Many human cancer cells overproduce hydrogen peroxide. High levels (up to 0.5 nmol/hr/10^4 cells) of hydrogen peroxide are constitutively released from a wide range of human tumor cells. I believe that this makes the tumor cell more vulnerable to increases in EMODs and creates selectivity for all modalities of cancer therapy. In other words, adding a specified amount of EMODs to neoplastic and normal cells can induce apoptosis in a cancer cell without doing harm to normal cells.

That aspect alone is the Holy Grail in cancer therapy!

RMH Note: Oxygen free radicals are designed to provide for the flow of electrons and I believe that some antioxidants are designed as possible co-oxidants or pre-oxidants (not oxidant scavengers or quenchers) in promotion of this electron/proton flow. Otherwise, one cannot explain the prooxidant activity of antioxidants.

Basic energy-producing metabolism for aerobic organisms starts with $^3\Sigma gO_2$ (ground state triplet oxygen) and ends with H_2O (water). The primary intermediate oxygen derivatives are $O_2^{\cdot-}$ (superoxide anion radical), H_2O_2 (hydrogen peroxide), the hydroxyl radical (.OH) and oxygen's primary electronic excitation state, $^1\Delta_g O_2$ (metastable singlet delta oxygen).

RMH 1-13-16
I now see a time or "Temporal Stratification of Reactivity" of the following EMODs: Long lived hydrogen peroxide; intermediate lived ozone, followed by superoxide anion and short lived singlet oxygen and the hydroxyl radical.

Both $O_2^{\cdot-}$ and H_2O_2 are essential steps in the mitochondrial electron transport chain and its coupling to oxidative phosphorylation for ATP (adenosine triphosphate) energy production. **All other EMODs are formed by secondary mechanisms**.

Antioxidants May Make Cancer Worse

Antioxidants May Make Cancer Worse. By Melinda Wenner Moyer on October 7, 2015. Scientific American

Antioxidants are supposed to keep your cells healthy. That is why millions of people gobble supplements like vitamin E and beta-carotene each year. Today, however, a new study adds to a growing body of research suggesting these supplements actually have a harmful effect in one serious disease: cancer.

The work, conducted in mice, shows that antioxidants can change cells in ways that fuel the spread of malignant melanoma—the most serious skin cancer—to different parts of the body. The progression makes the disease even more deadly. Earlier studies of antioxidant supplement use by people have also hinted at a cancer-promoting effect. A large trial reported in 1994 (pdf) that daily megadoses of the antioxidant beta-carotene increased the risk of lung cancer in male smokers by 18 percent and a 1996 trial was stopped early after researchers discovered that high-dose beta-carotene and retinol, another form of vitamin A, increased lung cancer risk by 28 percent in smokers and workers exposed to asbestos. More recently, a 2011 trial involving more than 35,500 men over 50 found that large doses of vitamin E increased the risk of prostate cancer by 17 percent. These findings had puzzled researchers because the conventional wisdom is that antioxidants should lower cancer risk by neutralizing cell-damaging, cancer-causing free radicals.

But scientists now think that antioxidants, at high enough levels, also protect cancer cells from these same free radicals. "There now exists a sizable quantity of data suggesting that antioxidants can help cancer cells much like they help normal cells," says Zachary Schafer, a biologist at the University of Notre Dame, who was not involved in the new study. Last year the scientists behind the melanoma study found that antioxidants fuel the growth of another type of malignancy, lung cancer.

For the new study, published in Science Translational Medicine, Martin Bergö, a cell biologist at the University of Gothenburg's Sahlgrenska Cancer Center in Sweden, and his colleagues decided to look at melanoma because rates have been increasing and because the cancer is known to be sensitive to the effects of free radicals. They fed the antioxidant N-acetylcysteine (NAC) to mice that had been genetically engineered to be susceptible to melanoma. The per-weight dose they gave the mice was consistent with what people typically consume in supplements. Although the treated mice did not develop more skin tumors than similar mice that had not been fed the antioxidants, they developed twice as many tumors in their lymph nodes, a hallmark of the spread of cancer—a process called metastasis. When the researchers added NAC or a form of vitamin E to cultured human melanoma cells, they confirmed that the antioxidants improved the cells' ability to move and invade a nearby membrane.

Antioxidants may bolster protection of these dangerous cells. Bergö and his colleagues found higher levels of glutathione, an antioxidant made by the body, inside metastatic tumor cells in treated mice compared with untreated mice. The treated mice also had a higher ratio of glutathione to glutathione disulfide, the molecule that glutathione becomes after it neutralizes free radicals. These findings suggest that when the body is given extra antioxidants, its tumor cells get to keep more of the antioxidants that they already make themselves. The cells can store the surplus, improving their ability to survive damage. This idea is supported by work that shows some genes that drive cancer growth turn on other genes that make intrinsic antioxidants.

The substances may help cancer cells in other ways, too. Previous research has suggested that glutathione affects the activity of a protein called RhoA, which helps cells move to different parts of the body. "If you were to select one protein that is known to be involved in [cell] migration, RhoA is it," Bergö explains. He and his colleagues confirmed that the extra glutathione in the treated mice caused levels of RhoA to increase in their metastatic tumors. In their 2014 lung cancer study they also found that antioxidant supplements caused lung tumor cells to turn off the activity of a well-known cancer-suppressing gene called p53; its inactivation is believed to drive metastasis. And Schafer's work has shown that antioxidants help migrating breast cancer cells survive when they detach from the extracellular matrix, the network of proteins surrounding cells.

These molecular investigations shed light on the large human trials that have implicated antioxidants in cancer. It is possible that the supplements did not trigger cancer but rather accelerated the progression of existing undiagnosed cancers, making later discovery of the disease likely. In other words, it "could be that while antioxidants might prevent DNA damage—and thus impede tumor initiation—once a tumor is established, antioxidants might facilitate the malignant behavior of cancer cells," Schafer says.

The medical advice for people at this point is tentative. More studies need to be done to bolster this hypothesis and understand exactly how antioxidants affect cancer cells in humans.

Bergö, who is not a medical doctor, does believe that people who are at an increased risk for lung cancer or melanoma or who have been diagnosed with either one should avoid antioxidant supplements. "There's no conclusive evidence that it would be beneficial to them, and there's mounting evidence that it could be harmful," he says.

His results do have a silver lining. They suggest a potential new way to target the disease. If cancer is very sensitive to the damaging effects of free radicals, then it might be possible to develop drugs that target cancer cells specifically and prevent them from producing antioxidants or that ramp up free radical levels inside of the malignant cells, exploiting their newly discovered weakness.

RMH Note: I have been saying this for decades.

Good Advice

Should patients be counseled against using antioxidant supplements to prevent or treat cancer? You bet they should!

I present approximately 106 studies showing EMODs are responsible for cancer cell killing (induction of apoptosis).

I present approximately 50 studies showing that antioxidants can block EMOD induced apoptosis.

I present one hundred eighteen (118) studies showing antioxidant ineffectiveness in cancer treatment or prevention and of this thirty-nine (39) studies showing harmful antioxidant effects (Taken from book Two, *Death in Small Doses?* and from *Antioxidant Overkill*) (Howes, 2010) (Howes, Overkill, 2011).

In my book, *Antioxidants Links to Deadly Unintended Consequences*, I presented over 500 study reports showing the ineffectiveness of antioxidants and of these, 170 study reports showed their harmful nature (Howes, 2012).

Consumption of excessive antioxidants to individuals who do not need them is unwise and can be harmful. Please read my books, *Death in Small Doses,* and *Antioxidant Overkill* (Howes, 2010).

Only ingest specific naturally occurring antioxidants if there is a proven antioxidant deficiency. Otherwise, there is a likelihood that you will be interfering with the specific intent of your tumoricidal therapy and of your overall health.

A 2005 report by D'Andrea concluded that cancer patients should avoid antioxidant supplements while receiving chemotherapy or radiation treatment.

Also, **a Wall Street Journal article argued that antioxidants could block the beneficial effects of standard cancer therapy** (Parker-Pope, 2005).

A 2008 article, in the ***Journal of the National Cancer Institute,*** published a review of randomized trial data, which suggested that **cancer patients should avoid the routine use of antioxidant supplements because they may potentially decrease the efficacy of radiation therapy by protecting the tumor and reducing overall survival** (Lawenda et al, 2008).

The concept is fairly simple: increased antioxidants (which decrease cancer cellular killing) decrease EMODs and apoptosis; whereas, decreased antioxidants increase EMOD affectivity and increase cancer cellular killing. It can be intentionally confused but it is basically that straight forward in systems relying on EMOD cytotoxicity and subsequent tumor apoptosis.

The *Harvard Science Review*, Spring 2010, published an article entitled, "***Antioxidants not heaven-sent.***" They cited my 2006 paper on the Free Radical Fantasy and stated, "It may come as a surprise that the current scientific consensus is that there is no health benefit to taking antioxidant supplements."

In February 2011, *Newsweek* magazine published an article entitled, "***Antioxidants Fall from Grace.***" Also saying, "The popular dietary components may not do any good and may actually harm."

In 2016, I formulated the following rules:

HOWES' NEW RULES

Following my exhaustive studies of the world literature and my selective reviews on oxygen metabolism, I have compiled these new rules for cancer patients.

Howes' Rules for Antioxidant Use in Cancer Patients or Patients with Premalignant Conditions:

Howes Rule One: Any medicine or supplement that is taken during cancer radiotherapy, chemotherapy or photodynamic therapy should have been tested to prove that it does not protect or promote neoplastic cells, because safeguarding tumor cells would defeat or negate the explicit intended effect and purpose of the anti-cancer (tumoricidal) treatment or therapy.

Howes Rule Two: Any reduction in a therapy's cancer-killing ability is in direct conflict with the overall medical treatment objective and the patients' ultimate wellbeing and should therefore be avoided.

Howes Rule Three: Based on significant prior data, excessive antioxidant levels theoretically, and in fact, promote an environment that is beneficial to tumor cell growth and survival. Considerable scientific evidence has shown that antioxidants can serve as shields for tumorous cells. Patients with known premalignant conditions or cancer (including survivors) should avoid antioxidant over use until further research shows otherwise.

Howes Rule Four: Even if an antioxidant is shown to reduce adverse effects of primary cancer treatments, it must be proven not to diminish the tumoricidal activity of the anti-cancer therapy. If it decreases the tumoricidal intent, in any way, it must be used with extreme caution, if at all and the patient must be informed of its possible harmful effect.

I strongly recommend against the use of unneeded antioxidants by cancer survivors, until further research yields a clear and unequivocal conclusion. After all, **"Why risk defeating the purpose of your cancer treatment, even by a small amount, by ingesting unnecessary antioxidants?"** The occurrence of adverse events could negate any potential benefits of treatments.

**Oxidants and antioxidants
sing in the choir of life.
It is a two-part harmony.
When their voices blend,
it is a beautiful melody of wellbeing,
homeostasis and oxidative bliss.**
R.M. Howes, M.D., Ph.D.
12-11-10

**Radicophobes can ignore the truth or
they can reject the truth but
they cannot change the magnificent truths
regarding the crucial role
of EMODs in the life process of all aerobes
or the inherent splendor of oxygen.**

R.M. Howes, M.D., Ph.D.

1-26-09

Spontaneous Regression of Cancer

Because most cancers that are detected are also treated, **there are only a few reports documenting spontaneous regression of breast cancer.** (Dussan et al, 2008) (Krutchik et al, 1978).

However, **spontaneous regression of advanced cancer has long been recognized in metastatic melanoma and metastatic renal cell carcinoma**, and, in fact, such observations have motivated the interest in immunotherapy in these settings. (Printz 2001) (Gleave et al, 1998) (de Gast et al, 2000).

Furthermore, **more systematic investigations of spontaneous regression are beginning to be reported in the context of screen-detected abnormalities**. There are data suggesting that **regression routinely occurs in colonic adenomas** (both from the National Polyp Study and others) (Loeve et al, 2004) (Hofstad et al, 1996).

There is a growing literature **documenting regression in precancerous lesions of the cervix.** (Schlecht et al, 2003) (Moscicki et al, 2004).

Documentation of regression in screen-detected cancer is limited to neuroblastoma, for which investigators have found that screening detects far more cancer than will ever become clinically apparent and that a substantial proportion regress. (Schilling et al, 2002) (Yamamoto et al, 1998).

Spontaneous Regression Has Been Seen In:
- **metastatic melanoma**
- **metastatic renal cell carcinoma**
- **precancerous lesions of the cervix**
- **neuroblastoma**
- **some prostate cancer**
- **some lung cancer**

Conclusions: Because the cumulative incidence among controls never reached that of the screened group, **it appears that some breast cancers detected by repeated mammographic screening would not persist to be detectable by a single mammogram at the end of 6 years.** This raises the possibility that **the natural course of some screen-detected invasive breast cancers is to spontaneously regress**. (Abstract: November 24, 2008 edition of the *Archives of Internal Medicine the Natural History of Invasive Breast Cancers Detected by Screening Mammography.* Per-Henrik Zahl, Jan Moehlen, H. Gilbert Welch).

Oxygen doesn't make you age.
It "allows" you to age.
Want proof?
Try aging without it!
R.M. Howes, M.D., Ph.D.
8-17-11

Some cancer cells are more sensitive to generated reactive oxygen species, and this may be a useful difference that can be exploited when seeking to kill cancer cells but spare normal cells.

Apoptosis itself is largely based on EMODs (free radicals) released from mitochondria. ROS generation in mitochondria no longer appears to be **a precise mechanism used in signaling pathways such as apoptosis**, as demonstrated in **cardiomyocytes** (Duranteau et al, 1998).

In most forms of cell suicide, **the signaling cascade utilizes reactive oxygen species as essential intermediate messenger molecules** (Albright, C. D., 2003) (Vrablic, A. S., 2001) (Slater, A. F., 1995) (Johnson, T. M., 1996) (Sugiyama, H., 1996).

Even modest quenching of oxygen radicals by dietary antioxidants could block completion of apoptosis. Administration of antioxidants subsequent to a mutagenic event may effectively intercept free radicals that are critical in promoting apoptosis.

My first clue about our ability to naturally kill cancer was related to "spontaneous regression or remission," which is defined as cancers that shrink or disappear completely, without any conventional or alternative anticancer therapy. It is well-established in the medical literature (hundreds of descriptions) and amongst oncologists.

The most frequently cited cancers that may experience a spontaneous remission include kidney and testicular cancers as well as lymphoma and melanoma (estimated at 1 out of every 400 cases). Perhaps the best studied group is a type of lymphoma referred to as "low-grade, B-cell, Non-Hodgkin's lymphoma." It is well established that approximately 20% of patients diagnosed with this type of lymphoma will experience a spontaneous shrinkage of their disease. It is for this reason that oncologists do not treat these types of lymphoma unless they are causing bothersome symptoms for the patient.

In the *New York Times* (Oct 27, 2009), Gina Kolata wrote an article entitled, "Cancers Can Vanish Without Treatment, but How?" The impetus for her article was recent medical publications on screening tests for breast cancer and prostate cancer (mammography and PSA testing). Kolata wrote, "Besides finding tumors that would be lethal if left untreated, screening appears to be finding many small tumors that would not be a problem if they were left alone, undiscovered by screening. They were destined to stop growing on their own or shrink, or even, at least in the case of some breast cancers, disappear."

I believe that EMODs are, at least to a significant degree, the basis of this phenomenon. They are primarily responsible for EMOD-induced apoptosis (cell suicide), along with a vast complexity of biochemical entities.

Please remember that EMODs can be blocked by antioxidants, which can interact when combined in what is called a "cocktail effect." Others refer to it as a "synergism." Actually, my work shows that combining the antioxidant vitamins increases their harmful consequences, which I presented and discussed in *Death in Small Doses?*

The developed nations have the highest rates of common cancers, such as prostate, breast and lung and it appears that we are transferring this to the lesser developed nations. **Antioxidants are in hundreds of food products and manufactured goods in the industrialized nations.** Are they causal of these diseases? Are they related to the BPAs (bisphenols) used in the plastics industry and which is an antioxidant used to prevent polymerization? We now have excellent leads to answers concerning antioxidants.

Are developed nations like the wealthy Romans, who could afford lead-laden plates and pots and subsequently died of lead poisoning, because they could afford it? The poor were spared, and this may be analogous to under developed nations not being able to afford the harmful products (antioxidants) of the industrialized nations (contaminated baby bottles, drink cans, water lines, fortified foods, antioxidant vitamin supplements, etc.).

There are those who say that "cancer is a pathway disease" and I believe that EMODs will lead the way to the cure and prevention of cancer. Redox (oxidation/reduction) pathways will lead us to the door of discovery and I am trying to open and go through that door.

67 Antioxidant Studies Linked to Increased Cancer Risk - A Distillation in Chronological Order: 1986-2014

1986

1. Studies on antioxidants: Their carcinogenic and modifying effects on chemical carcinogenesis (Ito et al, 1986). *Squamous-cell carcinomas were induced in the forestomach of rats and hamsters fed BHA* (butylated hydroxyanisole). **BHA and other antioxidants, particularly propyl gallate and ethoxyquin, showed additive effects in inducing forestomach hyperplasia and cytotoxicity. BHA enhanced forestomach carcinogenesis initiated in rats by *N*-Methyl-*N'*-nitro-*N*-nitrosoguanidine or *N*-methylnitrosourea (MNU) and enhanced urinary bladder carcinogenesis initiated by MNU or *N*-Butyl-*N*-(4-hydroxybutyl) nitrosamine (BBN)** (Ito et al, 1986).

2. Studies on antioxidants: Their carcinogenic and modifying effects on chemical carcinogenesis (Ito et al, 1986). BHT (butylated hydroxytoluene**) promoted urinary bladder carcinogenesis initiated by BBN (*N*-butyl-*N*-(4-hydroxybutyl) nitrosamine) or MNU and thyroid carcinogenesis initiated by MNU (*N*-methylnitrosouea). Ethoxyquin promoted EHEN-initiated kidney carcinogenesis** (Ito et al, 1986).

3. Pathology of BHA- and BHT-induced lesions (Moch, 1986). Papilloma and squamous-cell carcinoma of the forestomach were increased at the 2.0% level with BHA in dogs. The BHT was fed to Wistar rats at 0, 25, 100 and 250 mg/kg body weight. At the highest dose there was an increase in the number of rats with hepatocellular adenoma and with hepatocellular carcinoma (Moch, 1986).

1994

4. α-Tocopherol, β-Carotene Cancer Prevention Study (ATBC study) (Heinonen et al, 1994) (#29,133 men); *50% increase in hemorrhagic stroke deaths among vitamin E group; 11% increase in ischemic heart disease deaths among β-carotene group; 18% increase in lung cancer among β-carotene group.*

5. The β-Carotene and Retinol Efficacy Trial (CARET) (Omenn GS, Goodman GE, Thorquist MD, et al., 1996) (#14,254 heavy smokers and 4,060 asbestos workers) (total #18,314 men and women); *28% increase in lung cancer; 26% increase in CVD (nonsignificant); 17% increase in total mortality among treatment group. This study was stopped 21 months earlier than planned.*

6. Energy, nutrient intake and prostate cancer risk: a population-based case-control study in Sweden. (Andersson et al. 1996) (#1,062) *In age-adjusted analyses, there were positive associations of prostate cancer (all stages combined) risk with total energy intake as well as intake of total fat (saturated and monounsaturated), protein, retinol and zinc.* **The positive association with energy intake was stronger for advanced cancer, with an excess risk of 70% for the highest quartile vs. the lowest.** *After adjustment for energy intake, there was no apparent association of prostate cancers (all stages combined) with any of the investigated nutrients. However, a weak positive association between intake of retinol and advanced prostate cancer was observed.*

7. Alpha-Tocopherol and beta-carotene supplements and lung cancer incidence in the alpha-tocopherol, beta-carotene cancer prevention study: effects of base-line characteristics and study compliance. (Albanes et al, 1996) (#29,133 men, smokers) *beta-Carotene supplementation was associated with increased lung cancer risk. beta-Carotene supplementation at pharmacologic levels may modestly increase lung cancer incidence in cigarette smokers*, and this effect may be associated with heavier smoking and higher alcohol intake.

8. Inhibition of thyroid peroxidase by dietary flavonoids (Divi, Doerge, 1996).

Thyroid peroxidase (TPO) is the enzyme that catalyzes thyroid hormone biosynthesis. Most flavonoids tested were potent inhibitors of TPO. *These inhibitory mechanisms for flavonoids are consistent with the antithyroid effects observed in experimental animals and, further, predict differences in hazards for antithyroid effects in humans consuming dietary flavonoids. Chronic consumption of flavonoids, especially suicide substrates, could play a role in the etiology of thyroid cancer* (Divi, Doerge, 1996).

1997

9. Chemoprevention of aerodigestive cancer (Berwick, Schantz, 1997) Large-scale trials of the anti-oxidant beta carotene have been *disappointing*; they have shown that among heavy smokers and possibly heavy alcohol consumers, beta carotene increases risk for lung cancer incidence and mortality (Berwick, Schantz, 1997).

1998

10. The Nurses' Health Study and Folic Acid and Colon Cancer (Giovannucci et al, 1998) (#88,756 women taking vitamin C and B-carotene, for 8 years); Dr. Andy Ness, of Bristol University, reported in the British Medical Journal in Dec. 2004, that there is *the possibility of increased risk of breast cancer in women taking folic acid supplements throughout pregnancy.* The researchers followed up **2,928 pregnant women** who had taken part in a supplemental trial in the 1960s. *The risk of death from breast cancer was much higher in women who had received high doses of the supplement* than in those who had been given a placebo.

11. Estrogenic effects of genistein on the growth of estrogen receptor-positive human breast cancer (MCF-7) cells in vitro and in vivo (Hsieh et al, 1998).
Tumors were larger in the genistein (750 ppm)-treated group than they were in the negative control group, demonstrating that dietary genistein was able to enhance the growth of MCF-7 cell tumors in vivo. In summary, *genistein can act as an estrogen agonist in vivo and in vitro, resulting in the proliferation of cultured human breast cancer cells (MCF-7) and the induction of pS2 gene expression. Dietary genistein stimulates mammary gland growth and enhances the growth of MCF-7 cell tumors in ovariectomized athymic mice* (Hsieh et al, 1998).

12. Effects of soy-protein supplementation on epithelial proliferation in the histologically normal human breast. (McMichael-Phillips et al, 1998) *Supplementation studies conducted in humans have raised the possibility that soy may promote breast cancer development. In a* *trial that randomized women with benign or malignant breast disease to soy supplementation (60 g of soy containing 45 mg of isoflavones) or their normal diet daily for 2 wk, women receiving the soy supplements had increased serum genistein levels in comparison to women on a standard diet, and their histologically normal breast tissue exhibited enhanced breast epithelial cell proliferation and significantly increased progesterone receptor levels* (McMichael-Phillips et al, 1998).

1999

13. Antioxidant vitamin supplements: update of their potential benefits and possible risks. (Maxwell, 1999) **A number of long term, prospective, randomized, placebo-controlled trials examining the protective effect of antioxidant supplements have now been completed. Their results have been generally disappointing and have provided little evidence of efficacy.** *Of greater concern, they have unexpectedly raised concerns that antioxidants, notably beta carotene, might increase the rate of development of cancers in high risk individuals.* **For this reason, regular consumption of antioxidant vitamins supplements cannot yet be advocated as a healthy lifestyle trait** (Maxwell, 1999).

2000

14. Multivitamin use and mortality in a large prospective study. (Watkins et al, 2000) (#1,063,023 adults) **Multivitamin users had heart disease and cerebrovascular disease mortality risks similar to those of nonusers, whereas combination users had mortality risks that were 15% lower than those of nonusers.** Multivitamin and combination use had minimal effect on cancer mortality overall, *although mortality from all cancers combined was increased among male current smokers who used multivitamins alone or in combination with vitamin A, C, or E.*

15. *Genistein exposure of L5178Y mouse lymphoma cells at concentrations less than 100 nM induced micronuclei formation and mutagenesis at the thymidine kinase locus.* (Boos, Stopper, 2000).

2001

16. Randomized Trial of Supplemental ß-Carotene to Prevent Second Head and Neck Cancer (Mayne et al, 2001) (#264 patients who had been curatively treated for a recent early-stage squamous cell carcinoma of the oral cavity, pharynx, or larynx.); **Supplemental ß-carotene had no significant effect on second head and neck cancer or lung cancer.** Whereas none of the effects were statistically significant, the *point estimates suggested a possible decrease in second head and neck cancer risk but a possible increase in lung cancer risk.*

17. Mega-dose vitamins and minerals in the treatment of non-metastatic breast cancer: an historical cohort study (Lesperance et al, 2002) (#90 patients with non-metastatic breast cancer who received conventional treatment) *Breast cancer–specific survival (i.e., patients censored only at death from breast cancer) and disease-free survival were shorter in the nutrient-supplemented group* **than in the non-supplemented group, but the differences were not statistically significant.**

Investigators stated that, "**It is troubling that both** (Lesperance et al, 2002 and Ferreira et al, 2004) **reported results suggesting poorer survival with concurrent administration of antioxidants and cytotoxic therapy.**"

2002

18. Selenium and vitamin E supplements for prostate cancer: evidence or embellishment? (Moyad et al. 2002) (# not available) Selenium supplements provided a benefit only for those individuals who had lower levels of baseline plasma selenium. *Other subjects, with normal or higher selenium levels, did not benefit and may have an increased risk for prostate cancer. Vitamin E supplements in higher doses (> or =100 IU) were also associated with a higher risk of aggressive or fatal prostate cancer in nonsmokers from a past prospective study.*

2003

19. Vitamins E & A fail to reduce incidence or mortality of lung cancer: Cochrane Database Syst Rev. 2003. (Caraballoso et al., 2003) (#109,394 participants); *When beta-carotene was combined with retinol, data from a single study showed that there was a statistically significant*, **increased risk of lung cancer incidence and mortality** *in people with risk factors for lung cancer who took both vitamins A and E.*

20. Neoplastic and Antineoplastic Effects of Beta Carotene on Colorectal Adenoma Recurrence: Results of a Randomized Trial (Baron et al, 2003) (#864 subjects who had had an adenoma removed and were polyp-free); *For participants who smoked cigarettes and also drank more than one alcoholic drink per day, beta carotene doubled the risk of adenoma recurrence.*

21. Selenium supplementation and secondary prevention of nonmelanoma skin cancer in a randomized trial. (Duffield-Lillico, 2003) (#1,312). *selenium supplementation was associated with statistically significantly elevated risk of squamous cell carcinoma* and of total nonmelanoma skin cancer. **Results from the Nutritional Prevention of Cancer Trial conducted among individuals at high risk of nonmelanoma skin cancer continue to demonstrate that selenium supplementation is ineffective at preventing basal cell carcinoma and that** *it statistically significantly increases the risk of basal cell, squamous cell carcinoma and total nonmelanoma skin cancer.*

22. Lycopene increases urokinase receptor and fails to inhibit growth or connexin expression in a metastatically passaged prostate cancer cell line: a brief communication (Forbes et al, 2003) *Lycopene has a potentially unwanted effect of upregulating expression of the urokinase plasminogen activator receptor and facilitating invasion while failing to significantly inhibit proliferation or to induce detectable levels of the gap junctional protein connexin 43 expression.* Our results indicate that **some caution should be taken with regard to use of lycopene to treat potentially advanced and metastatic prostate cancers** (Forbes et al, 2003).

23. Food and Chemical Toxicology; Effects of Dietary Antioxidants and 2-Amino-3-Methylimidazo[4,5-f]-Quinoline on Preneoplastic Lesions and on Oxidative Damage, Hormonal Status and Detoxification Capacity in the Rat. (Breinholt et al. 2003) Using a rat model, researchers studied the effects of long-term administration of moderate to large amounts of the food-based **antioxidants lycopene, quercetin and resveratrol** given individually. ***The antioxidant supplements caused precancerous sores in the rats' livers as well as damage to the DNA of certain cells of their immune system*** (Breinholt et al. 2003).

2004

24. Dietary tenistein results in larger MNU-induced, estrogen-dependent mammary tumors following ovariectomy of Sprague-Dawley rats (Allred et al, 2004) ***Genistein at 750 p.p.m. increased the weight of estrogen-dependent adenocarcinomas in ovariectomized rats compared with the negative-control animals.*** Genistein also ***enhanced mammary gland growth and tumor development of estrogen-dependent cells when ovariectomized mice were treated with the chemical carcinogen 1-methyl-1-nitrosourea to induce mammary tumorigenesis. Genistein treatment also resulted in a higher percentage of proliferative cells in tumors and increased uterine weights when compared with negative-control animals. In an endogenous estrogen environment similar to that of a postmenopausal woman, dietary genistein can stimulate the growth of a mammary carcinogen MNU-induced estrogen-dependent mammary tumors*** (Allred et al, 2004).

2005

25. Use of multivitamins and prostate cancer mortality in a large cohort of US men. (Stevens et al, 2005) (#475,726 men who were cancer-free) ***Regular multivitamin use was associated with a small increase in prostate cancer death rates.***

26. A randomized trial of antioxidant vitamins to prevent second primary cancers in head and neck cancer patients (Bairati et al, 2005 Apr 6) (#540 patients with stage I or II head and neck cancer treated by radiation therapy) Compared with patients receiving placebo, ***patients receiving alpha-tocopherol supplements had a higher rate of second primary cancers during the supplementation period*** but a lower rate after supplementation was discontinued. Similarly, **the rate of having a recurrence or second primary cancer was higher during** but lower after supplementation with alpha-tocopherol. CONCLUSIONS: ***alpha-Tocopherol supplementation produced unexpected adverse effects on the occurrence of second primary cancers and on cancer-free survival.*** **Note:** ***Patients taking an antioxidant were 1.65 times more likely to suffer a return of their original cancer during the three years they were on the supplement. The risk was highest among those taking only vitamin E (1.86 times higher).***

27. Randomized trial of antioxidant vitamins to prevent acute adverse effects of radiation therapy in head and neck cancer patients (Bairati et al, 2005 Aug 20) (#540 patients with stage I or II head and neck cancer treated by radiation therapy) During the course of the trial, ***supplementation with beta-carotene was discontinued because of ethical concerns.*** **Quality of life was not improved by the supplementation.** ***The rate of local recurrence of the head and neck tumor tended to be higher in the supplement arm of the trial.*** **Note:** Researchers were concerned to find that ***the rate of local recurrence (that is, a return of the original cancer) was 54 percent higher among patients on the combination pill than those on placebo.*** There was a smaller but still worrisome increase among those on vitamin E only. ***This trial suggests that use of high doses of antioxidants as adjuvant therapy might compromise radiation treatment efficacy.***

28. Low-dose dietary phytoestrogen abrogates tamoxifen-associated mammary tumor prevention. (Lui et al, 2005) *In vitro studies of human and mouse* **mammary tumor** *cell lines confirm that low doses of genistein, co-administered with tamoxifen, promote cell proliferation.* In summary, **low-dose dietary** *isoflavones abrogated* **tamoxifen-associated mammary tumor prevention** *in vivo. Genistein attenuates the anti-tumor activities of tamoxifen* (Lui et al, 2005).

2006

29. Smoking, alcohol drinking, green tea consumption and the risk of esophageal cancer in Japanese men. (Ishikawa et al, 2006) (#9,008 men in Cohort 1 and 17,715 men in Cohort 2) *Cigarette smoking, alcohol drinking and green tea consumption were significantly associated with an increased risk of esophageal cancer. The population attributable fractions of esophageal cancer incidence that was attributable to smoking, alcohol drinking and green tea consumption were 72.0%, 48.6%, and 22.1%, respectively.*
CONCLUSIONS: Among the variables studied, *smoking has the largest public health impact on esophageal cancer incidence in Japanese men, followed by alcohol drinking and green tea drinking.*

2007

30. Health Professionals Follow-up Study (2007): Effect of vitamins C, E, A and carotenoids and the occurrence of oral pre-malignant lesions (Maserejian et al, 2007) (#42,340 men enrolled in the Health Professionals Follow-up Study) (#207 found with oral premalignant lesions); *A trend for increased risk of oral pre-malignant lesions was observed with vitamin E, especially among current smokers and with vitamin E supplements. Beta-carotene also increased the risk among current smokers.* However, **dietary vitamin C was significantly associated with a reduced risk of oral premalignant lesions.**

31. Antioxidant Supplementation Increases the Risk of Skin Cancers in Women but Not in Men. (Hercberg et al, 2007) (#French adults, 7,876 women and 5,141 men. Total # = 13,017) daily capsule of antioxidants (120 mg vitamin C, 30 mg vitamin E, 6 mg β-carotene, 100 μg selenium, and 20 mg zinc) or a matching placebo. *In women, the incidence of SC was higher in the antioxidant group.* **Conversely, in men, incidence did not differ between the 2 treatment groups.** Despite the small number of events, *the incidence of melanoma was also higher in the antioxidant group for women.*

32. National Institutes of Health State-of-the-Science Conference Statement: Multivitamin/Mineral Supplements and Chronic Disease Prevention (NIH State-of-the Science Panel. 2007). reports from RCTs that noted excess lung cancer occurring in asbestos workers and smokers consuming β-carotene. In addition, *esophageal cancer excess was found with long-term follow-up of older Chinese patients (the Linxian study by Blot et al.) treated with selenium, β-carotene, and vitamin E supplements* (Blot et al, 1993) (NIH State-of-the Science Panel. 2007).

33. Multivitamin use and risk of prostate cancer in the National Institutes of Health-AARP Diet and Health Study (Lawson et al, 2007) (#295,344 men) *Investigators found an increased risk of advanced and fatal prostate cancers among men reporting excessive use of* *multivitamins (more than seven times per week) when compared with never users. The positive associations with excessive multivitamin use were strongest in men with a family history of prostate cancer or who took individual micronutrient supplements, including selenium, beta-carotene, or zinc* (Lawson et al, 2007).

2008

34. Systematic review: primary and secondary prevention of gastrointestinal cancers with antioxidant supplements. (Bjelakovic, Nikolova, Simonette and Gludd, 2008 Sept) (#211,818 participants) *Antioxidant supplements had no significant effect on mortality in a random-effects model meta-analysis but significantly increased mortality in a fixed-effect model meta-analysis. CONCLUSIONS: There was no evidence that the studied antioxidant supplements prevented gastrointestinal cancers. On the contrary, they seem to increase overall mortality.*

35. VITAL (VITamins and Lifestyle) study (2008) (Slatore et al, 2008) (#77,721 men and women); *Supplemental vitamin E was associated with a small increased risk of lung cancer.*

36. Efficacy of Antioxidant Supplementation in Reducing Primary Cancer Incidence and Mortality: Systematic Review and Meta-analysis (Bardia et al, 2008) (#104,196 participants) *Beta carotene supplementation was associated with an increase in the incidence of cancer among smokers and with a trend toward increased cancer mortality.*

37. Continuous in vitro exposure to low-dose genistein induces genomic instability in breast epithelial cells. (Kim et al, 2008) *Genistein exposure of L5178Y mouse lymphoma cells at concentrations less than 100 nM induced micronuclei formation and mutagenesis at the thymidine kinase locus* (Boos, Stopper, 2000) (Boos, G., Stopper, H., Genotoxicity of several clinically used topoisomerase II inhibitors. Toxicol. Lett. 2000, 116, 7–16) *and chronic exposure (3 months) of human MCF-10A cells induced genomic instability* (Kim et al, 2008).

38. Dietary genistein negates the inhibitory effect of letrozole on the growth of aromatase-expressing estrogen-dependent human breast cancer cells. (Ju et al, 2008) *Dietary GEN increased the growth of MCF-7Ca tumors implanted in ovariectomized mice and could also negate the inhibitory effect of LET on MCF-7Ca tumor growth. GEN can reverse the inhibitory effect of LET on tumor growth and adversely impact breast cancer therapy.* Overall, **the data suggest that short-term exposure to soy caused a weak estrogenic effect.** *An additional concern about the potential harm that might be done by genistein stems from data that it can stimulate proliferation of tamoxifen-sensitive cells* (Ju et al, 2008).

39. Phytoestrogens and breast cancer: a complex story. (Helferich et al, 2008) *Dietary genistein was able to negate the inhibitory effect of TAM on E-stimulated tumor growth.* In summary, *genistein can act as an estrogen agonist resulting in proliferation of E-dependent human breast cancer tumors in vivo.* Additionally, dietary genistein can negate the inhibitory effects of TAM on E-stimulated growth of MCF-7 cell tumors implanted into ovariectomized athymic mice (Helferich et al, 2008).

40. The role of antioxidant supplement in immune system, neoplastic, and neurodegenerative disorders: a point of view for an assessment of the risk/benefit profile (Brambilla et al, 2008) **Although evidence shows that antioxidant treatment results in cytoprotection, the potential clinical benefit deriving from both nutritional and supplemental antioxidants is still under wide debate.** In this line, *the inappropriate assumption of some lipophilic vitamins has been associated with increased incidence of cancer rather than with beneficial effects* (Brambilla et al, 2008).

41. Why have antioxidants failed? Research does point to some potential concerns with antioxidant supplementation; for example, *beta-carotene supplements may increase the risk of lung cancer in smokers, and vitamin E supplements may increase the risk of bleeding in certain individuals* (Steinhubl, 2008).

2009

42. Effect of selenium and vitamin E on risk of prostate cancer and other cancers: The Selenium and Vitamin E Cancer Prevention Trial (SELECT) (2009) (Lippman et al, 2009, SELECT) (#35,533 men) **There were statistically nonsignificant increased risks of prostate cancer in the vitamin E group** but not in the selenium + vitamin E group. CONCLUSION: **Selenium or vitamin E, alone or in combination at the doses and formulations used, did not prevent prostate cancer in this population of relatively healthy men.** *The trial was stopped ahead of its original 12-year deadline because of a lack of any noticeable benefit.* **In 2011, after three more years of follow-up, researchers found that the men who had taken vitamin E (400 IU a day of the synthetic form) had a 17 percent increased risk of prostate cancer, compared to those taking a placebo; this was published in the** *Journal of the American Medical Association.* **This suggests that effects of supplements can show up years after people stop taking them.**

43. Total and Cancer Mortality After Supplementation with Vitamins and Minerals: 10-year Follow-Up of the Linxian General Population Nutrition Intervention Trial. (Qiao et al, 2009) (#29,584 adult participants) *esophageal cancer deaths increased 14% among those aged 55 years or older. Vitamin A and zinc supplementation was associated with increased total and stroke mortality.*

44. Long-term use of beta-carotene, retinol, lycopene, and lutein supplements and lung cancer risk: results from the VITamins and Lifestyle (VITAL) study. (Satia et al, 2009) (#77,126 (VITAL) *Longer duration of use of individual beta-carotene, retinol, and lutein supplements (but not total 10-year average dose) was associated with statistically significantly elevated risk of total lung cancer and histologic cell types.*

45. Vitamin and mineral use and risk of prostate cancer: the case-control surveillance study. (Zhang et al. 2009) (#1,706 prostate cancer cases and 2,404 matched controls). *Men who used zinc for ten years or more, either in a multivitamin or as a supplement, had an approximately two-fold increased risk of prostate cancer. The finding that long-term zinc intake from multivitamins or single supplements was associated with a doubling in risk of prostate cancer adds to the growing evidence for an unfavorable effect of zinc on prostate cancer carcinogenesis.*

46. Folic acid and risk of prostate cancer: results from a randomized clinical trial. (Figueiredo et al, 2009) (643 randomly assigned men). *the estimated probability of being diagnosed with prostate cancer over a 10-year period was 9.7% in the folic acid group and 3.3% in the placebo group.*

47. Green tea consumption and risk of stomach cancer: a meta-analysis of epidemiologic studies (Myung, Int J Cancer. et al, 2009) (#13 epidemiologic studies) *In the meta-analyses of the recent cohort studies, the highest green tea consumption was shown to significantly increase stomach cancer risk using the crude data,* **but no significant association between them was seen when using the adjusted data.**

48. Green tea (Camellia sinensis) for the prevention of cancer. Cochrane Database Syst Rev. 2009 Jul 8;(3):CD005004. (Boehm et al, 2009) (#Fifty-one studies with more than 1.6 million participants were included) **there was limited to moderate evidence that the consumption of green tea reduced the risk of lung cancer, especially in men, and** *urinary bladder cancer or that it could even increase the risk of the latter.*

49. In vitro evaluation of selenium genotoxic, cytotoxic, and protective effects: a review (Valdiglesias et al, 2009) *There exists a considerable literature indicating both cytotoxic and genotoxic effects of selenium when cells in culture or animals are provided high doses, and consequences include enhanced mutagenesis, induction of chromosomal abnormalities such as micronuclei formation, as well as induction of cell-cycle arrest and apoptosis* (Valdiglesias et al, 2009).

50. Researchers investigated the association between tea drinking habits in Golestan province, northern Iran and risk of esophageal squamous cell carcinoma. There were 300 histologically proved cases of esophageal squamous cell carcinoma and 571 matched controls in the case-control study and 48,582 participants in the cohort study. *Compared with drinking luke warm or warm tea, drinking hot tea or very hot tea was strongly associated with an increased risk of esophageal cancer* (Islami et al, 2009).

2010

51. Multivitamin use and breast cancer incidence in a prospective cohort of Swedish women. (Larsson et al, 2010) (#35,329 cancer-free women) *Multivitamin use was associated with a statistically significant increased risk of breast cancer. Use of multivitamins was linked to a statistically significant 19 per cent increased risk of breast cancer* (after adjusting for lifestyle and risk factors like weight, diet, smoking, exercise, and family history of breast cancer.

52. Quercetin and Ferulic Acid Aggravate Renal Carcinoma in Long-Term Diabetic Victims. (Chiu-Lan Hsieh et al, 2010). Conclusively, the phytoantioxidants *quercetin and ferulic acid are able to aggravate, if not induce, nephrocarcinoma in mice.*

53. Selenomethionine and alpha-tocopherol do not inhibit prostate carcinogenesis in the testosterone plus estradiol-treated NBL rat model (Ozten et al, 2010) *Alpha-Tocopherol significantly increased the incidence of adenocarcinomas of the mammary glands at both* *dietary concentrations in rats.* Importantly, the results of the current animal studies and those reported previously were fully predictive of the outcome of the Selenium and Vitamin E Cancer Prevention Trial (Ozten et al, 2010).

54. Source-specific effects of micronutrients in lung cancer prevention (Roswall et al, 2010, Lung cancer) (#55,557) evaluate the association between **vitamin C, E, folate and beta-carotene and lung cancer** risk. **We found a significant protective effect of dietary vitamin E intake** *and a significantly higher lung cancer risk with supplemental beta-carotene and dietary folate intak.* (Roswall et al, 2010, Lung cancer)

55. Incidence of skin cancers during 5-year follow-up after stopping antioxidant vitamins and mineral supplementation (Ezzedine et al, 2010) (#12,741) In the SU.VI.MAX study, antioxidant supplementation for 7.5 years was found to increase skin cancer risk in women but not in men. **randomized trial,** (daily a placebo or a combination of ascorbic acid (120 mg), vitamin E (30 mg), β-carotene (6 mg), selenium (100 μg) and zinc (20mg), from inclusion in 1994 to September 2002) *The risk of skin cancers associated with antioxidant intake declines following interruption of supplementation. This supports a causative role for antioxidants in the evolution of skin cancer.* (Ezzedine et al, 2010).

2011

56. The protective effects of nutritional antioxidant therapy on Ehrlich solid tumor-bearing mice depend on the type of antioxidant therapy chosen: histology, genotoxicity and hematology evaluations. (Miranda-Vilela AL, et al, 2011) **Antioxidant administrations before tumor inoculation effectively inhibited its growth in the three experimental protocols,** but *administrations after the tumor's appearance accelerated tumor growth and favored metastases.* **Continuous administration of pequi oil inhibited the tumor's growth, while the same protocol with** *vitamins E and C accelerated it (tumor growth), favoring metastasis and increasing oxidative stress on erythrocytes.*

57. Prenatal exposure to flavonoids: Implication for cancer risk (Vanhees et al, 2011). *In vitro exposure to genistein/quercetin induced higher numbers of Mll rearrangements in bone marrow cells of Atm-ΔSRI mutant mice compared with wt mice. Prenatal exposure to flavonoids associated with higher frequencies of Mll rearrangements and a slight increase in the incidence of malignancies in DNA repair-deficient mice.* These data suggest that *prenatal exposure to both genistein and quercetin supplements could increase the risk on Mll rearrangements especially in the presence of compromised DNA repair.*

58. Use of vitamin supplements and risk of total cancer and cardiovascular disease among the Japanese general population: a population-based survey (Hara et al, 2011) (#28,903 men and 33,726 women for a total of 62,629) In women, consistent vitamin supplement use was associated with lower risk of CVD, *whereas past and recent use of vitamin supplements were associated with higher risk of cancer* (Hara et al, 2011).

59. Antioxidants tied to mixed effects in breast cancer. (Greenlee, Hershman et al, 2009) *Women who regularly took a mix of carotenoids had a higher risk of dying from breast* *cancer. Dietary supplements containing high doses carotenoids may be harmful, and people should think twice before taking them"* (Greenlee, Hershman et al, 2009).

60. Vitamin E pills linked with prostate cancer risk. (Klein et al, JAMA, 2011) Men randomly assigned to take a 400-unit capsule of vitamin E every day for about five years were 17 percent more likely to get prostate cancer than those given dummy pills.

61. The combined influence of multiple sex and growth hormones on risk of postmenopausal breast cancer: a nested case-control study (Tworoger et al, 2011) (#265 cases and 541 controls in the prospective Nurses' Health Study) *Women in the top versus bottom quintile of individual estrogen or androgen levels had approximately a doubling of postmenopausal breast cancer risk (both are antioxidants). Multiple hormones with high circulating levels substantially increase the risk of breast cancer, particularly ER-positive disease.* **Elevated estrogens had the biggest effect on cancer risk, especially for ER-positive cancer** (Tworoger et al, 2011).

2012

62. The *Midspan Collaborative study found* that **men who drank more than seven cups of tea per day had a significantly increased risk of prostate cancer compared to those who drank no tea or less than four cups per day.** (The *Midspan Collaborative study.* Nutrition and Cancer, 2012, Kashif Shafique Glasgow University) Male tea drinkers 'may be at greater risk of prostate cancer' (Shafique et al, 2012).

2013

63. Evidence of harmful effect of bisphenol A-based plastics. **Academic studies indicate that the substance may increase the risk of cardiovascular diseases, breast and prostate cancer** as well as neuronal diseases. Various organizations have pointed out that bisphenol A may be hazardous to health: The Federal Institute for Risk Assessment (Bundesinstitut für Risikoforschung), the European Food Safety Authority, the US Food and Drug Administration (FDA), the US National Institutes of Health (NIH) and the US-American Breast Cancer Foundation. However, those organizations have not yet provided a final assessment of the substance's hazardous potential. Nevertheless, the European Commission banned the use of bisphenol A in the manufacture of baby bottles in 2011 (Schopel et al, 2013).

2014

64. January 29, 2014 in Science Translational Medicine, Martin Bergö and Per Lindahl, wrote **"Antioxidants NAC, Including Vitamin E Can Promote Lung Cancer."** A study in mice has found that two commonly used antioxidants — vitamin E and a compound called *N*-acetylcysteine (NAC) — speed the growth of lung cancer rather than curb it. The results, published today in *Science Translational Medicine*, may be contrary to the expectations of the millions of people who gulp down vitamin E supplements each day. A survey published in 2005[2] found that about 11% of US adults took vitamin E supplements, at doses of 400 international units (IU) or higher. The recommended daily intake is 22.4 IU. The researchers fed either NAC or vitamin E to the mice, using doses of 5 or 50 times higher than the daily recommended amount for mice. The results for the two antioxidants were similar: **tumors grew about three times faster than those in animals that did not receive the treatment. Treated mice also died from their cancers about twice as quickly as untreated mice.** *The antioxidants caused a 2.8-fold increase in lung tumors, made the tumors more invasive and aggressive, and caused the mice to die twice as quickly - all compared to mice not given antioxidants. When the antioxidants were added to human lung tumor cells in lab dishes, they also accelerated cancer growth.* **These antioxidants actually increased cancer burden and mortality in a dose-dependent manner** (Sayin et al, 2014).

65. In 2014, the United States Preventive Services Task Force (USPSTF) _recommended against the use of beta-carotene or vitamin E supplements for the primary prevention of cardiovascular disease or cancer_, **according to a recommendation statement being published in** *Annals of Internal Medicine*. **A meta-analysis of 27 studies that covered more than 450,000 participants and found that multivitamins had no beneficial effect on preventing cardiovascular disease or cancer.** In addition, **taking vitamins didn't prevent mortality in any way. However, the analysis did confirm that smokers who took only beta carotene supplements increased their risk of lung cancer. (see Reuters Health article of February 24, 2014, here: http://reut.rs/1fCKqCP).**

66. This analysis included an examination of whether that baseline selenium level had any impact on the development of prostate cancer. Those men who had received no supplementation had no change in the development of prostate cancer over time. However, _those men who were taking selenium and who had a high level of selenium to begin with had a significant, nearly 2-fold increase in the development of high-risk prostate cancer. Those men taking vitamin E, even those with a low level of selenium in their toenails, had a significant increase in the risk of developing any prostate cancer and high Gleason score prostate cancer_. This randomized study demonstrated **after 2 years of follow-up, the men who took vitamin E had a statistically significant 17% increased risk for prostate cancer.** *In the men with low levels of selenium randomized to receive vitamin E alone, the total risk for prostate cancer increased by 63% and the risk for high-grade cancer increased by 111%.* **"Men using these supplements should stop, period. Neither selenium nor vitamin E supplementation confer any known [health] benefits — only risks,"** said lead author Alan Kristal, Dr. PH, from the Fred Hutchinson Cancer Research Center in Seattle (Frankel et al, 2014).

67. Soy could adversely affect gene expression in breast cancer. There are conflicting reports on the impact of soy on breast carcinogenesis. This study examines the effects of soy supplementation on breast cancer-related genes and pathways. **Methods** Women (n = 140) with early-stage breast cancer were randomly assigned to soy protein supplementation (n = 70) or placebo (n = 70) for 7 to 30 days, from diagnosis until surgery. A high-genistein signature consisting of 126 differentially expressed genes was identified from microarray analysis of tumors. This signature was characterized by overexpression (>2-fold) of cell cycle transcripts, including those that promote cell proliferation, such as FGFR2, E2F5, BUB1, CCNB2, MYBL2, CDK1, and CDC20. *Gene expression associated with soy intake and high plasma genistein defines a signature characterized by overexpression of FGFR2 and genes that drive cell cycle* *and proliferation pathways*. **For some in the soy group, expression of a gene associated with cancer growth increase.** These findings raise the concerns that in a subset of women soy could adversely affect gene expression in breast cancer (Shike et al, 2014).

A List of 41 U.S. Food and Drug Administration-Approved <u>Nontargeted</u> Anticancer Drugs That Work Through Generation of Reactive Oxygen Species (EMODs):

Leucovorin; Cytarabine; Methotrexate; Fludarabine; Gemcitabine; Cytarabine; Capecitabine; 5-Fluorouracil; Clofarabine; Azacitidine; Nelarabine; Decitabine; Pralatrexate; Pemetrexed; Etoposide; Docetaxel; Paclitaxel; Ixabepilone; Cabazitaxel; Eribulin mesylate; Vincristine; Vinblastine; Bleomycin; Chlorambucil; Procarbazine; Dacarbazine; Ifosfamide; Temozolomide; Oxaliplatin; Bendamustine; Daunorubicin; Epirubicin; Doxorubicin; Irinotecan; Topotecan; Platinum Analogue; Cisplatin; Pegaspargase; Arsenic trioxide; Lenalidomide; Plerixafor.

A List of 40 U.S. Food and Drug Administration-Approved <u>Targeted</u> Anticancer Drugs That Work Through Generation of Reactive Oxygen Species (EMODs)

Rituximab; Ibritumomab tiuxetan; Tositumomab and I 131; Ofatumumab; Gemtuzumab ozagamicin; Alemtuzumab; Imatinib; Aldesleukin; Denileukin diftitox; Gefitinib; Panitumumab; Trastuzumab; Bevacizumab; Pazopanib; Lapatinib ditosylate; Vandetanib; Sorafenib tosylate; Sunitinib malate; Nilotinib; Dasatinib; Denosumab; Vorinostat; Romidepsin; Temsirolimus; Bortezomib; Ipilimumab; Plerixafor acetate; Leuprolide acetate; Abarelix; Degarelix; Anastrozole; Exemestane; Letrozole; Tamoxifen citrate; Toremifene; Raloxifene; Fulvestrant; Bexarotene.

RMH Note: This gives a total of 81 drugs that kill cancer and work through EMOD induced apoptosis.

In December of 1999, Salganik pointed out that, "**Almost all anticancer drugs kill cancer cells by way of apoptosis, and antioxidants like vitamin A and vitamin E dramatically reduce apoptosis in cancer cells." Patients should therefore avoid taking any more than a normal amount of these vitamins during chemotherapy treatment.** In short, do not add to normal intake but Dr. Salganik's advice went further indicating that "an antioxidant-depleted diet could improve cancer therapies."

As of 2008, Dr. Alison Ross, science information officer at Cancer Research UK, pointed out, **there is currently no evidence from clinical trials in humans that injecting or consuming vitamin C, or other antioxidants, is effective in treating cancer**. "**Some research even suggests that high doses of antioxidants can make cancer treatment less effective, reducing the benefits of radiotherapy and chemotherapy."**

My Summary of Cancer Cell Types Killed by EMODs:

H_2O_2 (an EMOD) causes apoptosis in:
hepatoma cells, leukemia cells, osteosarcoma, breast, bladder, and lung cancer cells. Also procarbazine produces peroxide that kills Hodgkin's lymphoma, non-Hodgkin's lymphoma and primary brain tumors.

ROS (EMODs) induced apoptosis in:
chronic myeloid leukemia, lymphoma cells, prostate cancer cells.

ROS (EMODs) induce autophagic death in:
breast cancer, non-small cell lung cancer (NSCLC), glioma, neuroblastoma, glioblastoma, and cervical cancer.

ROS (EMODs) generated by gemcitabine and by thymoquinone have also been shown to inhibit the growth of pancreatic and prostate cancer cells.

As2O3 (an EMOD generating agent) has the ability to induce superoxide production in cancer cells and treat acute promyelocytic leukemia.

Some cancer types for which ROS (EMODs) have been shown to play a role in radiation-induced cancer cell death are lung adeno carcinoma, nonsmall-cell-lung cancer, prostate cancer, and breast cancer.

Other clinical studies for which ROS (EMODs) have been shown to play a role in radiation-induced therapy include patients with head and neck squamous cell carcinoma, cervical cancer, prostate cancer, NSCLC, rectal cancer, and breast cancer.

RMH Conclusion: Despite intensive investigations in this field, current antioxidant therapeutics have not and are not clinically effective in combating cancer.

Moreover, if as Gupta insinuates, EMODs cause cancer, please remember that EMODs are ubiquitous and omnipresent. Thus, cancer would also be ubiquitous and omnipresent and it is not. If so, we would all be dead.

Randomized clinical human antioxidant trials have been a dismal failure for disease cure, reversal or prevention.

**Regarding fraudulent antioxidant vitamin claims,
you do not have to follow the science.
All you have to do is....
follow the money.**
R.M. Howes, M.D., Ph.D.
10-22-10

Heart Disease and Excess Antioxidants

In 2015, the WHO reported that approximately one third of global deaths were attributable to a cardiovascular disease (CVD)-related event. Atherosclerosis, an inflammatory disorder of the vasculature, is the primary cause of CVD-related events, including myocardial infarction (MI) and stroke.

Investigators of the subsequent GISSI-Prevenzione trial, published in 1999, recruited 11,324 patients with recent MI and randomly assigned them to receive omega-3 supplements (1g daily), vitamin E, both, or none for 3.5 years. The primary endpoint of the study was a composite of death, nonfatal MI, and stroke. After 6 months, **the study found no clinically important changes in the serum levels of total cholesterol, LDL-cholesterol, and HDL-cholesterol.** (Anon. 1999)

Despite the promising results discussed thus far, the benefits of omega-3 PUFA supplementation on cardiovascular health remain inconclusive, given the conflicting results in the literature. In 2014, a meta-analysis that included five trials enrolling 396 participants **found no significant reduction in CVD-related events in individuals with peripheral arterial disease.** (Enns J, et al., 2014)

A systematic review that included 48 randomized controlled trials (36,913 individuals) and 41 cohort studies also **did not detect any significant reductions in CVD-related mortality in patients receiving omega-3 supplementation for 6 months**. (Hooper L, et al., 2006)

A meta-analysis that specifically focused on patients with a history of CVD **was also unable to identify any substantial protective effects of omega-3 PUFA supplementation** in 14 randomized double-blind trials that recruited 20,485 participants. (Kwak S, Myung S, Lee Y, Seo H., 2012)

A further meta-analysis also failed to demonstrate any association between omega-3 supplementation and mortality risk after evaluation of 20 randomized clinical studies that included 68,680 individuals in total. (Rizos EC, Ntzani EE, Bika E, et al, 2012)

However, care must be taken when interpreting the results of clinical trials owing to the heterogeneity within the designs of the studies. All of these factors are likely to affect the outcomes of the trials, and result in the inconsistent results found within the clinical trials and meta-analysis.

Given that the human body is unable to store vitamin C (also known as ascorbic acid), it is vital that foods rich in vitamin C, such as oranges, orange juice, broccoli and blackcurrants, form part of the daily diet. Increased intake of vitamin C has long been associated with a decrease in the prevalence of coronary artery disease. (Osganian SK, et al., 2003)

Despite such promise, the use of vitamin C as a nutraceutical for the prevention of atherosclerosis remains controversial because **many studies have failed to show any benefit on plaque lesions or lipid profiles**.

Dietary supplementation with a cocktail of anti-oxidants (vitamin E, vitamin C, and β-carotene) in ApoE deficient mice **did not reduce lesion size or alter plasma lipid profile**. (Averill MM, et al., 2008)

Individual supplementation with vitamin C or E failed to reduce intima-media thickness, and the combined supplementation did not reduce atherosclerosis progression in women. (Salonen JT, et al., 2000)

Despite numerous positive findings, the inconsistencies in the results assessing vitamin C and vitamin E supplementation are also evident in many other clinical trials. A randomized study that used an initial 2g dose followed by a daily intake of 1g of vitamin C in 20 young adult smokers showed improved vasodilation after the first 2 hours, **but there were no sustained beneficial effects after 8 weeks**. (Raitakari OT, et al., 2000)

In addition, a large-scale study involving 20,536 adults in the UK with either coronary artery disease, peripheral occlusive arterial disease, or diabetes **that were randomly assigned a daily dietary supplement containing either vitamin E, vitamin C, β-carotene, or placebo reported no observable benefits in terms of all-cause mortality or CVD-events at the 5-year follow up**[157]. (Anon. MRC/BHF, 2002) (Anon. MRC/BHF Heart Protection Study of antioxidant vitamin supplementation in 20,536 high-risk individuals: a randomized placebo-controlled trial. Lancet. 2002; 360:23–33)

Furthermore, several studies have failed to demonstrate any cardiovascular protective effects following vitamin E consumption. The previously mentioned GISSI-Prevenzione trial found that daily consumption of vitamin E (300 mg) was not associated with a reduced risk of CVD-related events[50].

The HOPE study, which involved 9,541 participants considered to be at high risk of a CVD-related event, was **also unable to find any significant reductions in cardiovascular deaths following daily vitamin E consumption for 4.5 years**. (Yusuf S, Dagenais G, Pogue J, et al., 2000)

On the other hand, the VEAPS trial observed a decrease in plasma oxLDL levels and a reduction in the vulnerability of LDL to oxidation in 353 individuals following daily vitamin E supplementation for 3 years. (Hodis HN., 2002)

However, this trial also demonstrated that **vitamin E supplementation was unable to reduce the intima-media thickness compared to the placebo, indicating that it was unable to prevent atherosclerosis development**. (Hodis HN., 2002)

In conclusion, although both vitamin C and E were once considered ideal nutraceuticals for the prevention of atherosclerosis owing to their antioxidative and vasodilatory properties, **they have not been proven to be consistently effective in long-term prevention of CVD.**

This position is consistent with the AHA whose advisory panel in 2004 recommended against using vitamin supplements to reduce the risk of CVD-related events. (Kris-Etherton PM, Lichtenstein AH, Howard BV, et al., 2004)

Research has shown that vegetables such as Brussels sprouts and broccoli protect a person from developing cancer by increasing production of oxidants (Bjlakovic pp. 842-57 et al, 2007).

Mortality in randomized trials of antioxidant supplements for primary and secondary prevention: systematic review and meta-analysis. (Bjlakovic et al., pp. 765-6, 2008)

Bjelakovic G1, Nikolova D, Gluud LL, Simonetti RG, Gluud C.

JAMA. 2008 Feb 20;299(7):765-6.

Abstract

Antioxidant supplements are used for prevention of several diseases.

OBJECTIVE:

To assess the effect of antioxidant supplements on mortality in randomized primary and secondary prevention trials. DATA SOURCES AND TRIAL SELECTION: We searched electronic databases and bibliographies published by October 2005. All randomized trials involving adults comparing beta carotene, vitamin A, vitamin C (ascorbic acid), vitamin E, and selenium either singly or combined vs placebo or vs no intervention were included in our analysis. Randomization, blinding, and follow-up were considered markers of bias in the included trials. The effect of antioxidant supplements on all-cause mortality was analyzed with random-effects meta-analyses and reported as relative risk (RR) with 95% confidence intervals (CIs). Meta-regression was used to assess the effect of covariates across the trials.

DATA EXTRACTION:

We included 68 randomized trials with 232 606 participants (385 publications).

DATA SYNTHESIS:

When all low- and high-bias risk trials of antioxidant supplements were pooled together there was no significant effect on mortality (RR, 1.02; 95% CI, 0.98-1.06). **Multivariate meta-regression analyses showed that low-bias risk trials (RR, 1.16; 95% CI, 1.04[corrected]-1.29) and selenium (RR, 0.998; 95% CI, 0.997-0.9995) were significantly associated with mortality. In 47 low-bias trials with 180 938 participants, the antioxidant supplements significantly increased mortality** (RR, 1.05; 95% CI, 1.02-1.08). **In low-bias risk trials, after exclusion of selenium trials, beta carotene (RR, 1.07; 95% CI, 1.02-1.11), vitamin A (RR, 1.16; 95% CI, 1.10-1.24), and vitamin E (RR, 1.04; 95% CI, 1.01-1.07), singly or combined, significantly increased mortality**. Vitamin C and selenium had no significant effect on mortality.

CONCLUSIONS:

Treatment with beta carotene, vitamin A, and vitamin E may increase mortality. The potential roles of vitamin C and selenium on mortality need further study.

Antioxidant Profile and Early Outcome in Stroke Patients

(Cherubini A, Maria Cristina Polidori, Mario Bregnocchi, et al., 2000) (Cherubini A, Maria Cristina Polidori, Mario Bregnocchi, Salvatore Pezzuto, Roberta Cecchetti, Tiziana Ingegni, Angelo di Iorio, Umberto Senin, Patrizia Mecocci. Antioxidant Profile and Early Outcome in Stroke Patients. Stroke. 2000; 31:2295-2300)

Ischemic stroke is a leading cause of mortality and disability in Western countries, particularly in the elderly. Human studies on stroke and oxidative stress in the brain are still lacking.

Antioxidants have been evaluated as neuroprotective agents in stroke since there is evidence supporting the occurrence of a condition of oxidative stress in the brain during ischemia. In experimental studies, an increased free radical generation during cerebral ischemia/reperfusion injury has been shown in vivo using several techniques such as microdialysis, salicylate spin trapping, and electron paramagnetic resonance.

An increase of lipid peroxidation products and a decrease in tissue antioxidant levels in the brain during ischemia have been reported as indirect evidence of oxidative stress. Pharmacological studies in animals showed that antioxidant molecules able to cross the blood-brain barrier, such as polyethylene glycol–conjugated superoxide dismutase (SOD) and catalase and lazaroids, reduce ischemic cerebral damage.

Finally**, transgenic mice overexpressing SOD have reduced infarct size compared with wild-type mice, while SOD knockout mice have an increased infarct size compared with controls**.

RMH Note: This is because SOD produces beneficial levels of hydrogen peroxide from the superoxide anion.

Lower serum SOD activity has been found in acute stroke patients. Higher SOD levels would have been protective.

Recently, it has been shown that total peroxyl radical trapping potential of plasma as well as ascorbic acid, α-tocopherol, and protein thiol plasma levels are inversely correlated with neurological impairment after cerebral infarction.

In other studies, however, no differences in vitamin C or in vitamin A or E concentrations were found between stroke patients and controls.

In another study, the urinary excretion of F2-isoprostanes, a specific in vivo marker of free radical damage to lipids, was similar in acute ischemic stroke patients and controls.

Experimental studies performed **in transgenic mice overexpressing SOD and in rats treated with exogenous SOD showed that this enzyme plays a protective role toward cerebral damage** induced by ischemia.

More recently, extracellular SOD activity was found to be lower in acute stroke patients but increased to the level of controls within 5 days from the cerebrovascular accident.

Although the regulation of SOD by oxidative stress is not yet completely defined, it has been demonstrated in rat neurons that conditions that increase oxidative stress do not induce the enzyme production.

Vitamin C is a cofactor in several metabolic activities and represents the major water-soluble antioxidant in the human body. Vitamin C was consistently lower in patients who had the worst outcome.

RMH Note: It is well known that vitamin C produces hydrogen peroxide when present in high levels, as is seen in intravenous vitamin C cancer therapy with ascorbate. Low vitamin C levels argues that there is also a lower level of peroxide, which is needed by the brain.

In conclusion, this longitudinal study of antioxidant levels during the first week after acute ischemic stroke reveals that almost all antioxidants are reduced immediately after a cerebrovascular accident and increase over the following days, suggesting that **the high antioxidant levels were present and contributory to the stroke itself.**

T cells and reactive oxygen species (EMODs)

(Belikov AV, Schraven, Simeoni L, 2015) (Belikov AV, Schraven, Simeoni L. **T cells and reactive oxygen species.** *Journal of Biomedical Science* **201522:85**)

Abstract

Reactive oxygen species (ROS) **(EMODs)** have been long considered simply as harmful by-products of metabolism, which damage cellular proteins, lipids, and nucleic acids. ROS are also known as a weapon of phagocytes, employed against pathogens invading the host. However, during the last decade, an understanding has emerged that **ROS (EMODs) also have important roles as signaling messengers in a multitude of pathways, in all cells, tissues, and organs**. T lymphocytes are the key players of the adaptive immune response, which both coordinate other immune cells and destroy malignant and virus-infected cells. **ROS (EMODs) have been extensively implicated in T-cell hypo-responsiveness, apoptosis, and activation**. It has also become evident that the source, the kinetics, and the localization of ROS production all influence cell responses. Thus, the characterization of the precise mechanisms by which ROS are involved in the regulation of T-cell functions is important for our understanding of the immune response and for the development of new therapeutic treatments against immune-mediated diseases. This review summarizes the 30-year-long history of research on ROS in T lymphocytes, with the emphasis on the physiological roles of ROS.

Introduction to ROS (EMODs)

ROS **(EMODs)** are small short-lived oxygen-containing molecules that are chemically highly reactive, a property that is mainly due to their unpaired electrons (radicals). *Superoxide* ($O_2{}^{\bullet-}$), *hydrogen peroxide* (H_2O_2), hydroxyl radical ($OH^{\bullet}$), hypochlorous acid ($HOCl$), lipid peroxides ($ROOH$), singlet oxygen ($^1O_2$), and ozone (O_3) are some of the most common ROS. The first two species are the most important ROS involved in the regulation of biological processes. $O_2{}^{\bullet-}$ is usually the species from which other ROS originate. Once produced, $O_2{}^{\bullet-}$ either rapidly reacts with surrounding molecules or dismutates to H_2O_2, spontaneously or with the help of *superoxide dismutase* (SOD). H_2O_2 is more stable, less reactive, can diffuse in the microenvironment and even cross cell membranes. H_2O_2 can either react with particular amino acids, usually cysteines and methionines, or can be converted to $OH^{\bullet}$ (in the Fenton reaction), $HOCl$ (by myeloperoxidase), or H_2O (with the help of *catalase, peroxidase*, or *peroxiredoxin*). Both $OH^{\bullet}$ and $HOCl$ are highly reactive and usually irreversibly damage nearby molecules.

One of the major sources of ROS **(EMODs)** in the cell are mitochondria. Mitochondria express the *electron transport chain* (ETC.) complexes, which transfer electrons from NADH and succinate, along a controlled redox pathway, to the oxygen molecule (O_2). Upon receiving four electrons, O_2 is reduced to H_2O. However, the ETC is not perfect, and occasionally O_2 undergoes one- or two-electron reduction to form $O_2 \cdot^-$ or H_2O_2, respectively. *Complexes I and III* of the ETC are the main sources of mitochondrial $O_2 \cdot^-$.

Multiple metabolic enzymes, such as ERO-1, cytochromes P-450 and b5, lipoxygenases, cyclooxygenases, α-ketoglutarate- and glycerol phosphate dehydrogenases, as well as hydroxyacid-, urate-, xanthine-, monoamine-, diamine-, polyamine-, and amino acid oxidases, are also producing ROS **(EMODs)** as necessary intermediates or byproducts of their reactions. These enzymes can be found in mitochondria, endoplasmic reticulum, peroxisomes, and cytosol. There is also a large class of ROS producing enzymes called *NADPH oxidases* (see below).

Finally, there are exogenous sources of ROS **(EMODs)**, including ultraviolet and gamma radiation, smoke and other air pollutants, as well as several drugs and chemicals. As ROS can damage proteins, lipids, and nucleic acids, the evolution has created specialized antioxidant systems. There are antioxidant enzymes, such as SODs, catalases, *glutathione peroxidases* (GPXs), *peroxiredoxins* (PRXs), *thioredoxins* (TRXs), glutaredoxins (GRXs), sulfiredoxins (SRXs), thioredoxin reductases, glutathione reductases, and methionine sulfoxide reductases, and also small nonenzymatic antioxidant molecules, such as *glutathione*, *ascorbate*, pyruvate, α-ketoglutarate, and oxaloacetate. Hypothetically, when the rate of ROS production in the cell (or in the microenvironment) significantly exceeds the rate of their neutralization by the antioxidant systems, the cell undergoes *oxidative stress*. Theoretically, prolonged or excessive oxidative stress can lead to the impairment of cellular functions, cell death or senescence.

Phagocytic cells produce ROS **(EMODs)** to kill engulfed bacteria during the so-called *respiratory burst*. Indeed, during the last decade, it has become evident that ROS are not just harmful byproducts of metabolism and weapons of phagocytes but are also crucial players in cellular signaling. **ROS-mediated signaling is involved in multiple processes, such as cell growth, stem cell renewal, tumorigenesis, cell death], cell senescence, cell migration], oxygen sensing, angiogenesis, circadian rhythm maintenance, and immune responses.**

Among ROS (EMODs), H_2O_2 acts as the major signaling messenger and is excellently suited for this function. In fact, it is stable enough, is able to cross cell membranes, and is reacting preferentially with cysteine residues. It has to be noted that cysteine residues are amongst the most conserved and least abundant protein residues, which ensures high selectivity and specificity for oxidation-mediated post-translational modifications. Moreover, only specific cysteinyl thiols that, upon coordination with neighboring amino acid side chains, can become thiolate anions are able to react with H_2O_2. When H_2O_2 oxidizes a cysteine thiolate anion (R-S$^-$), sulfenic acid (R-SOH) is formed. This process, which is referred to as *sulfenylation*, is reversed by GRXs and TRXs. Thus, sulfenylation is believed to be akin to phosphorylation or other post-translational modifications. Sulfenylation may lead to further post-translational modifications, such as glutathionylation, disulfide bond formation, and sulfinilation. Most importantly, it can be involved in the regulation of protein activity. In fact, sulfenylation can induce changes in the protein conformation, thus leading to the activation or inactivation of the catalytic center or to other functional alterations of the protein. Multiple protein classes have been shown to be regulated by sulfenylation, including phosphatases and kinases, transcription factors and histone deacetylases, antioxidant enzymes and heat-shock proteins, proteases and hydrolases, ion channels and pumps, adaptor molecules and cytoskeleton components.

Moreover, due to the abundance of antioxidant systems in the cell, ROS **(EMODs)** cannot travel long distances, and hence they transmit signals only locally, in confined compartments. In other words, the source and the corresponding targets of signaling ROS usually have to be in close proximity. ROS-mediated signaling can be additionally regulated via controlled alterations in local levels and activity of specific antioxidants.

T Cells and Oxidative Stress

T cells are often present in close proximity to phagocytic cells, which are known to produce large amounts of ROS **(EMODs)** during respiratory burst. Moreover, activated T cells can trigger respiratory burst by direct contacts with phagocytes as well as by secreted cytokines.

Overall, under physiological conditions, there is likely an equilibrium between ROS **(EMODs)** and antioxidant systems, which is required for the proper functioning of T cells.

ROS (EMODs) as regulators of activation-induced cell death

The stimulation of previously activated T cells – also called *T-cell blasts* – results in *activation-induced cell death* (AICD), a process which is accompanied by the release of ROS. Moreover, this stimulation also induces transient DCFDA oxidation that is not dependent on Fas or NOX-2, as well as the Fas-dependent oxidation of the superoxide-sensitive dye *dihydroethidium* (DHE) that is not mediated by NOX-2, **indicating that at least two more sources of ROS (EMODs) are involved in AICD.**

In summary, it seems that several different sources of ROS are involved in the AICD of T cells. First, H_2O_2 produced by DUOX-1 upon TCR triggering serves to amplify proximal signaling events downstream of the TCR. Next, $O_2^{\cdot-}$ released from mitochondrial Complex I, potentially in response to ERK signaling, triggers the expression of FasL. Finally, Fas ligation activates NOX-2, which probably assists the execution of the apoptotic program via the H_2O_2-mediated activation of AKT and the inhibition of MEK. Moreover, cell-intrinsic antioxidants, such as glutathione, vitamin E, MnSOD, and CuZnSOD, interfere with FasL expression, thus counteracting AICD. A better understanding of the role of ROS in the mechanisms of AICD could help in the development of therapeutic strategies for diseases in which T cells either die excessively or are not properly cleared during the contraction phase of immune response, such as AIDS or autoimmune diseases, respectively.

$O_2^{\cdot-}$ production is completely abrogated in T-cell preparations from NOX-2-deficient mice.

Another approach that has been used to understand the functional importance of ROS (EMODs) in the activation of primary T cells is incubation with antioxidants. In murine T cells, *butylated hydroxyanisole* (BHA) blocks DNA synthesis and CD25 expression, whereas the *mitochondria-targeted vitamin E* (Mitovitamin E) abrogates IL-2 production. In human T cells, *nordihydroguaiaretic acid* (NDGA) inhibits IL-2 synthesis, whereas vitamin E suppresses IL-4 production. Interestingly, catalase, superoxide dismutase, and ascorbate do not affect the expression of CD25 and CD69, proliferation, and cytokine production in human T cells. It seems that antioxidants that inhibit lipid peroxidation and/or lipoxygenase activity (such as NDGA, BHA, Vitamin E, or Mitovitamin E), but not antioxidants that scavenge water-soluble ROS (such as catalase, superoxide dismutase, or ascorbate) can interfere with the activation of primary T cells.

Conclusions and Perspectives

T cells are not isolated entities and are present in various tissue microenvironments. **It has become evident that the surrounding cells create a particular redox milieu that may, in turn, influence T-cell responses.**

Additionally, **appropriate intracellular ROS (EMOD) levels in T cells are created by the controlled production of ROS (EMODs) via NOX-2, DUOX-1, and mitochondria and the expression of a variety of antioxidant systems, including superoxide dismutases, peroxiredoxins, and glutaredoxins.**

Indeed, the available data suggest that ROS (EMODs) play crucial roles in T-cell biology.

First, **ROS participate in activation-induced cell death and hence in the termination of the immune response. Several different sources of ROS appear to be involved in this process. It seems that DUOX-1 produces H_2O_2 that amplifies proximal TCR signaling, mitochondrial Complex I releases $O_2^{\cdot-}$ that triggers the expression of FasL, and NOX-2 generates H_2O_2 that assists the execution of the apoptotic program.**

Second, ROS **(EMODs)** appear to be involved in the differentiation of T cells but dispensable for T-cell activation or proliferation. However, **this delicate balance is disturbed in several diseases, such as systemic sclerosis, rheumatoid arthritis, AIDS, and cancer.**

https://jbiomedsci.biomedcentral.com/articles/10.1186/s12929-015-0194-3

Reactive oxygen species (EMODs) in cancer
(Liou G-Y, 2010) (Liou G-Y. **Reactive oxygen species in cancer. Free Radical Research. Volume 44, 2010 - Issue 5. Pages 479-496**)

Elevated rates of reactive oxygen species (ROS) **(EMODs)** have been detected in almost all cancers, where they promote many aspects of tumor development and progression. However, tumor cells also express increased levels of antioxidant proteins to detoxify from ROS **(EMODs)**, suggesting that a delicate balance of intracellular ROS levels is required for cancer cell function. Further, the radical generated, the location of its generation, as well as the local concentration is important for the cellular functions of ROS in cancer.

A challenge for novel therapeutic strategies will be the fine tuning of intracellular ROS signaling to effectively tip the balance towards ROS-induced apoptotic signaling.

THE NAKED MOLE RAT

The Naked mole-rat (*Heterocephalus glaber*)

Background Information

The naked mole-rat (Heterocephalus glaber) also known as the "sand puppy" or desert mole rat, is a burrowing rodent native to parts of East Africa and is the only species currently classified in the genus Heterocephalus. The naked mole-rat and the Damaraland mole-rat are the only known eusocial mammals. It has a highly unusual set of physical traits that enable it to thrive in an otherwise harsh underground environment; it is the only mammalianthermoconformer, has a lack of pain sensation in its skin, and has very low metabolic and respiratory rates. It is also remarkable for its resistance to cancer and its longevity.

Oxygen Utilization

The naked mole-rat is well adapted for the limited availability of oxygen within the tunnels that are its habitat: its lungs are very small and **its blood has a very strong affinity for oxygen, increasing the efficiency of oxygen uptake. It has a very low respiration and metabolic rate for an animal of its size, about 2/3 that of a similarly sized mouse, thus using oxygen minimally.** In long periods of hunger, such as a drought, its metabolic rate can be reduced by up to 25 percent.

Thermoregulation

The naked mole-rat does not regulate its body temperature in typical mammalian fashion. They are thermoconformers rather than thermoregulators in that, unlike other mammals, body temperature tracks ambient temperatures. However, it has also been claimed that "the Naked Mole-Rat has a distinct temperature and activity rhythm that is not coupled to environmental conditions." The relationship between oxygen consumption and ambient temperature switches from a typical poikilothermic pattern to a homeothermic mode when temperature is at 28 °C or higher.

Cancer Resistance

Naked mole-rats appear to have a high resistance to tumors; cancer has never been observed in them. A potential mechanism that averts cancer is an "over-crowding" gene, p16, which prevents cell division once individual cells come into contact (known as "contact inhibition"). The cells of most mammals, including naked mole-rats, undergo contact inhibition via the gene p27 which prevents cellular reproduction at a much higher cell density than p16 does. The combination of p16 and p27 in naked mole-rat cells is a double barrier to uncontrolled cell proliferation, one of the hallmarks of cancer.

On June 19, 2013, scientists reported that the reason naked mole-rats do not get cancer may be because they produce an "extremely high-molecular-masshyaluronan" (HMW-HA) (a natural sugary substance), which is over "five times larger" than that in cancer-prone humans and cancer-susceptible laboratory animals.

The breakthrough scientific report was published a month later as the cover story of the journal Nature. A few months later, the same University of Rochester research team announced that **naked mole-rats have ribosomes that produce extremely error-free proteins. Because of both of these discoveries, the journal Science named the naked mole-rat "Vertebrate of the Year" for 2013.**

Blind mole-rats Spalax golani and Spalax judaei also appear to be immune to cancer but by a different mechanism.

Lifespan

The naked mole-rat is also of interest because it is extraordinarily long-lived for a rodent of its size (up to 31 years) and holds the record for the longest living rodent. Naked mole-rats are highly resistant to cancer and maintain healthy vascular function longer in their lifespan than shorter-living rats. The reason for their longevity is debated, but is thought to be related to their ability to substantially reduce their metabolism during hard times, and so prevent aging-induced damage from oxidative stress. This has been referred to as "living their life in pulses". Their longevity has also been attributed to "protein stability." Because of their extraordinary longevity, an international effort was put into place to sequence the genome of the naked mole-rat. **A draft genome was made available in 2011 with an improved version released in 2014.** Further transcriptome sequencing revealed genes related to mitochondria and oxidation reduction processes to have high expression levels in the naked mole-rat when compared to mice, which may contribute to their longevity.

Habitat

The naked mole-rat is native to the drier parts of the tropical grasslands of East Africa, predominantly southern Ethiopia, Kenya, and Somalia.

Clusters averaging 75 to 80 individuals live together in complex systems of burrows in arid African deserts. The tunnel systems built by naked mole-rats can stretch up to three to five kilometres (2–3 mi) in cumulative length.

The life-history of naked mole rats is astonishingly similar to social insects, with a dominant queen, populating the colony with son and daughter slaves.

Naked mole rats have well developed eyes, but their brain has lost the ability to process what they see, leaving them functionally blind.

Rodriguez et al. specifically focused on the longest-lived rodent, **the naked mole-rat, which maintains good health and provides novel insights into the paradox of maintaining both an extended health span and lifespan despite high oxidative stress from a young age**. (Rodriguez et al, 2011)

The toxic nature of oxygen was already a well-known phenomenon since the seminal work of Lavoisier in 1781. (Lavosier, 1781)

However, **free radicals were first regarded as the cause of oxygen toxicity in 1954 and soon afterward in 1956 Denham Harman proposed that physiological metals would cause** reactive oxygen species (ROS) (EMODs) to form in cells potentially damaging nearby molecules, including DNA. (Gershman et al, 1954) (Harman, 1956)

These EMODs would allegedly cause mutations, and based on the belief at the time, such induction of mutations could cause both cancer and aging.

RMH Note: The naked mole rat, backed up by the salutary effects of exercise in humans, have proved Harman and Ames wrong on both counts, i.e., aging and cancer.

Harman also proposed that administering compounds that could oxidize easily and absorb the ROS (EMODs) in the cell could slow down this mutation-induced aging.

Since that time, the free radical theory of aging has been repeatedly modified and renamed to the "oxidative stress theory of aging. (Beckman, Ames, 1998)

While a shortened lifespan may not be the product of accelerated aging, determining the lifespan of animal or plant species with compromised antioxidant pathways can be used further to explore this theory: it follows that <u>if an organism has increased oxidative damage but exhibits no change in lifespan, the result falsifies the hypothesis of the FRT and the MFRTA</u>.

RMH Note: That is exactly what happens in the naked mole rat, i.e., it has increased oxidative damage, yet, it lives 8 times as long as other similar rodents. (Buffenstein, 2008, CPB)

The naked mole rat is strictly subterranean, eusocial rodent, found in the northeast horn of Africa lives more than 30 years in captivity while maintaining cancer-free, good health well into its third decade of life. (Buffenstein, 2008, CPB)

The lack of spontaneous neoplasia is most unusual among captive wild-caught rodents (such as Mus musculus and Peromyscus species) as well as domesticated laboratory strains of mice and rats. Approximately 70% of domesticated laboratory rodent deaths are attributed to various types of cancers. (Ikeno et al, 2005)

Naked Mole-Rat Genome

Scientists sequenced DNA of a cancer-resistant rodent.
July 6, 2011

Scientists have generated the first whole-genome sequencing data of the naked mole-rat, a rodent that is resistant to cancer and lives for more than 30 years. The naked mole-rat is native to the deserts of East Africa and has unique physical traits that allow it to survive in harsh environments for many years.

The naked mole rat has a lack of pain sensation in its skin and has a low metabolic rate that allows it to live underground with limited oxygen supply.

Scientists at the University of Liverpool, in partnership with The Genome Analysis Centre, Norwich, have generated the first whole-genome sequencing data of the naked mole-rat.

For the first time, scientists have sequenced the genome of the naked mole-rat to understand its longevity and resistance to diseases of aging. Researchers will use the genomic information to study the mechanisms thought to protect against the causes of aging, such as DNA repair and genes associated with these processes.
To date, cancer has not been detected in the naked mole-rat.

Recent studies have suggested that **its cells possess anti-tumor capabilities that are not present in other rodents or in humans**.

In fact, **when naked mole rat cells are induced to form a tumor, the rodents stop the threat almost immediately**.

In announcing the Vertebrate of the Year honor, Science cited two research papers published this year, both written by Gorbunova and Seluanov. According to the magazine, one paper explained **how a ribosome in naked mole rats "excels at producing error-free proteins," while the other focused on "a supersized version of a complex sugar that...builds up in the spaces between cells and may keep the cells from clumping together and forming tumors."**

A sugar called hyaluronic acid oozes from cells in the NMR creature's connective tissue and forms cages around the cells to prevent them from replicating unchecked, as happens in tumors and this sugar activates the cluster of anti-cancer genes called INK4.

The low level of carbonylated proteins in the naked mole-rat nuclear lysates may, therefore, indirectly contribute to their pronounced resistance to cancer.

PROOF EMODs (ROS) ARE AN ESSENTIAL PART OF CANCER KILL

Oxygen is not optional.
It is essential for creating and
perpetuating aerobic life and for
sustaining homeostatic health,
especially in man.
R. M. Howes, M.D., Ph.D.
4-2-15

Radically increase your chances
for establishing long term homeostatic health
by increasing your oxidative capacity,
- your oxidative bliss.
R. M. Howes, M.D., Ph.D.
4-3-15

The antioxidants beta carotene, lycopene and omega-3 fatty acids all increase the risk of the aggressive form of prostate cancer. <u>More and more well-designed studies show that antioxidants, in supplement form, do not deliver promised health benefits</u>.

Linking EMOD levels, apoptosis and cancer proliferation

SOD mimetics kill cancer because they produce peroxide and antioxidants block cancer kill because they block EMODs, especially H_2O_2. This is a pattern I have seen repeated many times and convinces me that antioxidants are dangerous.

Howes' rational for EMOD benefits and antioxidant harm

One of the important functions of apoptosis is the elimination of preneoplastic and neoplastic cells. (Thompson, C. B., 1995).

Some cancer cells are more sensitive to generated reactive oxygen species, and this may be a useful difference that can be exploited when seeking to kill cancer cells but spare normal cells.

Apoptosis itself is largely based on EMODs (free radicals) released from mitochondria. ROS (EMOD) generation in mitochondria no longer appears to be **a precise mechanism used in signaling pathways such as apoptosis**, as demonstrated in **cardiomyocytes**. (Duranteau et al, 1998).

In most forms of cell suicide, **the signaling cascade utilizes reactive oxygen species as essential intermediate messenger molecules** (Albright, C. D., 2003) (Vrablic, A. S., 2001) (Slater, A. F., 1995) (Johnson, T. M., 1996) (Sugiyama, H., 1996). =
Even modest quenching of oxygen radicals by dietary antioxidants could block completion of apoptosis. Administration of antioxidants subsequent to a mutagenic event may effectively intercept free radicals that are critical in promoting apoptosis. ++

Apoptosis, necrosis, and growth arrest can all be regulated by ROS (EMODs). (Aw, 1999). **ROS (EMODs) can also induce cellular senescence and apoptosis and can therefore function as anti-tumorigenic species.** (Valko et al, 2006). =

Tumor cell anti-oxidant defenses. Inhibition of the glutathione redox cycle enhances macrophage-mediated cytolysis. (Nathan et al, vol 153, 1981)

The basis of resistance to oxidative injury was studied in **six murine tumor cell lines (TLX9, P388, YAC, P815, J774 and NK lyphoma cell lines)** that differed 54-fold in their resistance to enzymatically generated H_2O_2. **This clearly illustrates the role of antioxidants, such as glutathione or selenium, as a cancer cell defense or protective system.**

I believe that **just as with bacterial pathogens, cancer cells endeavor to protect themselves from death activation by EMODs and this is accomplished via antioxidants or antioxidant enzymes.**

EMODs - A Primary Means of Killing Cancer

Tumor cells have a complex array of genetic changes and develop an alteration in the metabolism of oxygen, having a higher steady state of EMOD production. The altered oxygen metabolism of cancer cells is not subject to the high genetic variability of tumors and serves as a point of selectivity in cancer therapy and protection of normal cells.

Cancer cells have high prooxidant levels and this presents a point of selective kill of the cancer cells by EMOD induction of apoptosis. In other words, if we add EMODs to cancer cells, we can get them to kill themselves, without having the same effect on normal cells. This whole concept is key to a future successful cancer eradication and prevention program.

With retinoblastoma, we have tissue which uses more oxygen that any other in the body on a per gram basis. Yet, it has an extremely rare occurrence rate of cancer, which is contrary to the teachings of the Free Radi-crap theory.

I believe that this is analogous to the fact that tumors are rare in the brain, heart, thyroid and the retina, even though they have the highest oxygen consumption levels, with high EMOD generation (and low antioxidant levels) in the body. Ergo, tumorigenesis is not related to EMOD levels.

Summary of Chemotherapeutic Agents Which Generate EMODs:

Oxygen dependence has also been established for a number of chemotherapeutic agents such as cyclophosphamide, carboplatin, doxorubicin, etc. and levels are different for each agent. Hypoxia can impart resistance to many chemotherapeutic agents (Teicher et al, 1990). The chemotherapeutic agents doxorubicin, mitomycin C, etoposide and cisplatin are superoxide generating agents. (Yokomizo et al, 1995).

Many antitumor agents, such as vinblastine, cisplatin, mitomycin C, doxorubicin, camptothecin, inostamycin, neocarzinostatin and many others exhibit antitumor activity via EMOD induced apoptotic (ROS-dependent apoptosis activation) cell death, suggesting use of EMODs as an antitumor principle. (Fang, Nakamura, Iyer, 2007). In fact, some experts now believe that all chemotherapeutic agents generate some level of apoptosis activating EMODs.

Chemotherapeutic agents can **act solely via the production of reactive oxygen species and induction of apoptosis** (Ratnam et al, 2006). These agents include the **anthracyclines (e.g., doxorubicin), platinum-containing complexes (e.g., cisplatin, carboplatin), alkylating agents (e.g., cyclophosphamide, ifosfamide), and cytotoxic antibiotics (e.g., bleomycin, mitomycin-C).**

Although some classes of antineoplastic agents generate high levels of oxidative stress, others, including the taxanes, vinca alkaloids, antifolates, and nucleoside and nucleotide analogues, generate only low levels. Nevertheless, all drugs generate some free radicals as they induce apoptosis in cancer cells.

Many antitumor agents, such as vinblastine, cisplatin, mitomycin C, doxorubicin, camptothecin, inostamycin, neocarzinostatin and many others exhibit antitumor activity via (EMOD) ROS-dependent activation of apoptotic cell death, suggesting potential use of ROS (EMODs) as an antitumor principle. (Fang, Nakamura, Iyer, 2007).

EMODs, antioxidants and apoptosis: all antineoplastic drugs yield EMODs

Many chemotherapeutic drugs have well-defined mechanisms of actions, including traditional alkylating agents and anthracycline antitumor antibiotics, which generate EMODs. **Depending upon specifics of oxidation/reduction potentials, these EMODs are uniformly subject to transformation to altered compounds by antioxidants through the simple process of electron transfer.**

Radiotherapy Produces Extensive Amounts of EMODs

In tissues exposed to treatment, **radiotherapy produces cytotoxic free radicals and extensive oxidative stress**. (Cook et al, 2004) (Doroshenko et al, 2004).

Ionizing radiation produces a number of reactive oxygen species, including superoxide ion, hydroxyl radical, and hydrogen peroxide; it is currently postulated that the primary therapeutic effects from radiotherapy are mediated through the generation of cytotoxic reactive oxygen species. (Cook et al, 2004) (Lee et al, 2004).

Furthermore, **cancer cells accumulate iron, and radiotherapy elevates intracellular free iron concentrations that can lead to the production of cytotoxic hydroxyl radicals through the Fenton reaction.** (Kwok, Richardson, 2002) (Weijl et al, 2004).

Therefore, **cells exposed to radiotherapy are subjected to heightened oxidative stress**.

Methods or therapies that provide increased oxygen to cancer cells help radiation work more effectively by enabling more EMOD or free-radical formation.

Radiation kills cancer cells by concentrating massive amounts of free radicals directly into tumors.

To be effective, oxygen must be present during radiation or at least during the lifetime of the free radical (10^{-5} s).
Without oxygen, the indirect radiation damage can be repaired and thiols, e.g., glutathione, or other antioxidants can compete with this oxidation.

Photodynamic Therapy (PDT)

Photodynamic Therapy (PDT) holds considerable promise in treating cancer but current terminology leads to confusion.

When generated under carefully controlled conditions using exogenous sensitizers and light in the visible range (400 -700 nm), **$^1O_2^*$ can be exploited for therapeutic purposes, as for example, in antineoplastic photodynamic therapy (PDT).**

Remember, that **all antibodies go through a singlet oxygen and ozone step.** Antibodies can generate hydrogen peroxide (H_2O_2) from singlet molecular oxygen ($^1O_2^*$). This process is

catalytic, and we identify the electron source for a **quasi-unlimited generation of H_2O_2. Antibodies produce up to 500 mole equivalents of H_2O_2 from $^1O_2*$, without a reduction in rate. This work shows the enormous potential for H_2O_2 production by antibodies.** (Wentworth et al, 2001) (Wentworth et al, 2002).

Epilogue

My Approach to Killing Cancer

For over four decades, I have studied the arcane science of oxygen metabolism. I have published (free on line at *www.iwillfindthecure.org*) a magnum opus of about 12,000 pages, with tens of thousands of scientific references, which corrects the myth that electronically modified oxygen derivatives (EMODs) are inherently destructive. In fact, they are normally of low toxicity and are absolutely essential for subcellular signaling and energy production (Howes, UTOPIA, 2004).

It is unprecedented to accuse an essential product of normal metabolism, in normal levels, to be toxic and/or lethal.

Moreover, I believe that an EMOD insufficiency is the basis of disease allowance with diseases such as cancer, atherosclerosis, diabetes, arthritis, HIV/AIDS, Alzheimer's disease and malaria.

An adequate EMOD level is also essential for controlling neoplastic or cancerous growths by the induction of cellular apoptosis (cellular suicide). **This feature offers a unique site for selectively killing cancerous cells and not injuring normal cells** (Howes, Philica. Feb 26, 2007).

Most doctors are not familiar with my approach to controlling cancer. They have not spent decades doing arduous research into the arcane biochemistry of cancer causation and prevention, as I have.

I will provide you with a deluge of valuable information to kill and suppress cancer cell growth, all of which is evidence-based scientific studies.

At times, I will have to get highly technical to keep critics at bay and to validate my prooxidant approach.

At other times, my book will appear to be repetitive, but this is due to the fact that I have to emphasize and re-emphasize my points to overcome six decades of mythology promoted by the invalidated free radical theory.

I will try to present the most understandable information about cancer, its underlying causes and what is necessary to counteract cancer.

We must defeat cancer and we must do it now. There may be well over 100+ types of cancer but I firmly believe that there are common underlying principles for all of them, especially as it relates to cancer killing, suppression and protection.

Thus, my prooxidative approach applies to all types of cancer, such as lung, prostate, breast, colon, ovarian, cervical, liver, pancreatic, bone, bladder, stomach, testicular, thyroid, kidney, throat, brain, mouth, uterine, esophageal, rectal and more.

The fundamental causes of these cancers are the same. So, the same biochemical strategies work for all of them.

My principles also likely primarily underlie the occurrence of spontaneous regression of cancer.

My goal has been to bring these prooxidative methods to the patient's bedside and at basically no cost to them.

Many types of cancer have long development periods but short periods of survival once diagnosis has been established.

Cancer therapy has to consider the trials and tribulations of chemotherapy and radiation therapy. Also, in advanced cancer cases, one may have to move slowly to accommodate a compromised immune system, lethargy, cachexia and catabolic wasting.

Prooxidant cancer therapy is amenable to a combinatorial approach and should be considered in most cases. Prooxidant therapy can serve as an adjunct to all forms of cancer therapy.

If an individual consumes excessive amounts of antioxidants (which block EMOD action), they can literally promote cancerous growth and spread. It is abundantly clear that in many instances, antioxidants shield and protect cancerous cells.

Please do not listen to the misleading approaches based on the debunked free radical theory. It will "radically" mislead you. If you fall victim to the fallacious practices of the free radical theory and ingest excessive levels of antioxidants, you will weaken your overall oxidative defensive system.

My approach emphasizes raising our overall oxidative capacity to a point of "oxidative bliss" and to lowering antioxidant levels to a point where they no longer protect or shield cancer cells from an oxidative death (apoptosis).

We rely on electronically modified oxygen derivatives, EMODs (formerly called reactive oxygen species, ROS, or oxygen free radicals), to protect us from pathogenic bacteria, fungi, protozoans, viruses and cancer.

It is that simple!

Prooxidative therapies are compatible with chemotherapy, radiation therapy and surgery. There is no reason to avoid it because it will only complement the other forms of cancer treatment.

Please remember that chemotherapy and radiation therapy can wipe out your immune system, speed up cancer's spread and destroy your oxidative ability to kill cancer. Thus, you must be willing to boost or restore it to a point of "oxidative bliss."

This is especially true for patients who have been told that there is nothing that can be done. That is just plain wrong!

Many studies have shown that additional oxygenation of cancerous cells increases the killing effectiveness of radiation therapy, and to a lesser extent, chemo.

Also, basically all chemo drugs generate EMODs and in doing so, they cause the cancer cells to kill themselves (apoptotic execution or suicide).

One of my fundamental findings is that low EMOD levels promote or "allow" cancer development and spread and high EMOD levels kill and prevent cancer formation.

Low EMOD levels can be associated with low oxygenation levels.

In 1931 Dr. Warburg won his first Nobel Prize for proving cancer is caused by a lack of oxygen respiration in cells. He stated in an article titled *The Prime Cause and Prevention of Cancer* that **"the cause of cancer is no longer a mystery, we know it occurs whenever any cell is denied 60% of its oxygen requirements."**

"Cancer, above all other diseases, has countless secondary causes. But, even for cancer, there is only one prime cause. Summarized in a few words (according to Warburg), the prime cause of cancer is the replacement of the respiration of oxygen in normal body cells by a fermentation of sugar. All normal body cells meet their energy needs by respiration of oxygen, whereas **cancer cells meet their energy needs in great part by fermentation**. All normal body cells are thus obligate aerobes, whereas all cancer cells are partial anaerobes."

That, he said, was the prime cause of cancer until his death in 1970.

Most students of natural health are familiar with Dr. Warburg's name, as he won the Nobel Prize for medicine in 1931 for his discovery that cancer cells have a fundamentally different energy metabolism compared to healthy cells.

But they don't know he was a personal friend of German physicists Albert Einstein and Max Planck and was awarded a second Nobel Prize in 1944 but Hitler prevented him from going to Stockholm to pick it up. He is considered by most experts to be the greatest biochemist of the 20th century.

Reducing cellular oxygen levels by just 35% causes cancer in most every case according to Otto Warburg's Nobel Prize winning research.

Nobel laureate, James Watson noticed this and is claimed to have said: "[I]f we're ever going to cure cancer, we're clearly going to have to go back to the days of Otto Warburg and focus on the metabolism (of oxygen) to make any real progress."

However, I have found that poor oxygenation is linked to physiological aging. As we get older, we get less oxygen to our cells. And, the older we get, the more we are prone to develop cancer. Coincidence? I think not!

The reason for the failed war against cancer stems from a flawed paradigm that categorizes cancer as an exclusively free radical disease. My approach lends the potential to treat all types of cancer because it exploits the one weakness that is common to every cancer cell: an EMOD insufficiency that does not trigger their death threshold.

Decades ago, two researchers at the National Cancer Institute, Dean Burn and Mark Woods, (Dean translated some of Warburg's speeches) conducted a series of experiments where they measured the fermentation rate of cancers that grew at different speeds. What they found supported Dr. Warburg's theory. <u>The cancers with the highest growth rates had the highest fermentation rates (lowest oxygen levels)</u>. The slower a cancer grew, the less it used fermentation to produce energy.

Even in those very slow growing cancer cells, fermentation was still taking place, albeit at very low levels.

Pietro Gullino, also at the National Cancer Institute, devised a test which showed that this slow growing cancer *always* produced fermentation lactic acid (an antioxidant).

Silvio Fiala, a biochemist from the University of Southern California, also confirmed that this slow growing cancer produced lactic acid (an antioxidant), and that it's oxygen respiration was reduced.

Further research into Warburg's theory showed that when oxygen levels were turned down, cells began to produce energy anaerobically. **They ultimately became cancerous when levels went low enough, and it took a reduction of 35% in oxygen levels for this to happen.**

Pete Pedersen, Ph.D. at Johns Hopkins took Dr. Warburg's theory to the next level, morphologically determining that **there's a radically reduced number of mitochondria in cancer cells**.

J. B. Kizer, a biochemist and physicist at Gungnir Research in Portsmith, Ohio explained, "Since Warburg's discovery, this difference in respiration has remained the most fundamental (and some say, only) physiological difference consistently found between normal and cancer cells. Using cell culture studies, Kizer decided to examine the differential responses of normal and cancer cells to changes in the oxygen environment.

"The results that Kizer found were rather remarkable. Kizer found that... "**High O_2 tensions were lethal to cancer tissue,** 95 percent being very toxic, whereas in general, normal tissues were not harmed by high oxygen tensions. Indeed, some normal tissues were found to require high O_2 tensions. It does seem to demonstrate the possibility that if the O_2 tensions in cancer tissues can be elevated, then the **cancer tissue may be able to be killed selectively**, as it seems that the cancer cells are incapable of handling the O_2 in a high O_2 environment."

Cancer cells produce excess lactic acid, an antioxidant, as they ferment energy. Lactic acid is toxic and tends to prevent the transport of oxygen into neighboring normal cells and it counteracts EMODs. Over time, as these cells replicate, the cancer may spread if not destroyed by the immune system and other sources of EMODs.

I have found that, for their protection, cancer cells selectively concentrate antioxidants like lactic acid, vitamin C and glutathione, to protect themselves from an EMOD induced apoptotic death.

This is analogous to pathogenic bacteria evolving a catalase enzymatic system to protect and shield themselves from a hydrogen peroxide induced death.

Further, I believe that cancer cells have evolved to have lower numbers of mitochondria as a method of self-protection from an EMOD induced apoptotic death.

Researcher Tom Seyfried has done a remarkable job of compiling supporting evidence for the metabolic theory of cancer. For example, he dug up so-called nuclear transfer studies, most of which date back to the 1980s. They were very simple, elegant experiments in which they took the nucleus of a cancer cell and put it into a normal cell with its nucleus removed. The cells are then grown in a petri dish, after which they're injected into mice, to see what happens.

What they discovered was that when you take the nucleus of a cancer cell, put it in a normal cell, and put it in mice, *nothing* happens. No cancers develop, and the cells revert back to normal. This despite the fact that you have just inserted cells that have all the driving mutations purported to cause cancer! So why don't you get cancer?

At the time, all they could say was that something in the cytoplasm suppresses cancer. The experiment was then flipped, and when the nucleus of a normal cell was put into a cancer cell, which was then injected into mice, about 98 percent of the animals developed cancer. This is irrefutable evidence that something *in the cytoplasm* is not only repressing cancer, but is driving cancer too.

EMODs are produced primarily by the mitochondria, which reside in the cytoplasm. The above experiments show that the cancerous cells have an insufficiency of both mitochondria and consequently an insufficiency of EMODs. Thus, cancer is "allowed" to develop, flourish and metastasize.

It is in this state of "insufficiencies" that damage to the nuclear DNA occurs and mutations prevail.

No doubt, I have proven that high level EMODs are death to an incredibly wide spectrum of human and murine cancerous cell lines. Of additional concern, I have shown that the cancerous cells are shielded and protected from this EMOD induced death by excessive levels of antioxidants, especially N-acetyl cysteine, (NAC).

The implication of this research is that an effective way to support the body's fight against cancer would be to get as much **oxygen** as you can into healthy cells, and improving their ability to utilize oxygen, with the formation of EMODs. Raising the oxygen levels of normal cells would help prevent them from becoming cancerous.

And, increasing oxygen levels in cancer cells to high levels could help kill those cancer cells.

Ma Lan, MD and Joel Wallach DVD, point out that one type of white blood cells kills cancer cells by <u>injecting oxygen creating hydrogen peroxide</u> into the cells.

In 2018, there are reports that lactate serves as fuel for cancer cells. However, I believe that since it is an antioxidant, it more accurately serves to create an EMOD insufficiency and therefore prevents EMOD induced apoptosis and allows for cancer cell growth and metastasis.

Conclusion

The most commonly consumed vitamin and mineral supplements provide no consistent health benefit or harm, suggests a new study led by researchers at St. Michael's Hospital and the University of Toronto.

Published May 29, 2018 in the *Journal of the American College of Cardiology*, the systematic review found that multivitamins, vitamin D, calcium, and vitamin C—the most common supplements—showed no advantage or added risk in the prevention of cardiovascular disease, heart attack, stroke, or premature death. "We were surprised to find so few positive effects of the most common supplements that people consume," said Dr. David Jenkins, the study's lead author. "Our review found that if you want to use multivitamins, vitamin D, calcium, or vitamin C, it does no harm—but there is no apparent advantage either."

The study found folic acid alone and B vitamins with folic acid may reduce cardiovascular disease and stroke.

Meanwhile, niacin and antioxidants showed a very small effect that might signify an increased risk of death from any cause. As I said, "Taking excessive antioxidant supplements is tantamount to suicide."

I have worked relentlessly to teach my theories to my colleagues and have pushed to provide the benefits of my work to all patients at the lowest possible cost and in the safest, effective form. I bid you adieu.

**My arduous 75-year trek of discovery
to help mankind,
has yielded an expected
fulfilled sense of completion.
My phantasmagoric odyssey has been worth it!**
R.M. Howes, M.D., Ph.D.
6-5-18

REFERENCES

(Adams, R B; Egbo, K Nkechiyere; Demmig-Adams, Barbara., 2014) (Adams, Robert Benjamin; Egbo, Karen Nkechiyere; Demmig-Adams, Barbara. High-dose vitamin C supplements diminish the benefits of exercise in athletic training and disease prevention. Nutrition & Food Science, Volume 44, Number 2, 2014, pp. 95-101(7).

(Albanes D., Heinonen O.P., Taylor P.R., 1996) (Albanes D., Heinonen O.P., Taylor P.R., Virtamo J., Edwards B.K., Rautalahti M., Hartman A.M., Palmgren J., Freedman L.S., Haapakoski J., et al. Alpha-tocopherol and beta-carotene supplements and lung cancer incidence in the alpha-tocopherol, beta-carotene cancer prevention study: Effects of base-line characteristics and study compliance. J. Natl. Cancer Inst. 1996; 88:1560–1570).

(Albright et al, 2003) (Albright, C. D., Salganik, R. I., Craciunescu, C. N., Mar, M. H. & Zeisel, S. H. (2003) Mitochondrial and microsomal derived reactive oxygen species mediate apoptosis induced by transforming growth factor-beta1 in immortalized rat hepatocytes. J. Cell Biochem. 89:254-261).

(Algotar AM, Stratton MS, Ahmann FR, et al., 2013) (Algotar AM, Stratton MS, Ahmann FR, et al. Phase 3 clinical trial investigating the effect of selenium supplementation in men at high risk for prostate cancer. Prostate. 2013; 73:328–35).

(Allred et al, 2004) (Allred et al, Dietary genistein results in larger MNU-induced, estrogen-dependent mammary tumors following ovariectomy of Sprague-Dawley rats. Carcinogenesis. 2004 Feb;25(2):211-18).

(Andersson et al. 1996) (Andersson SO, Wolk A, Bergström R, Giovannucci E, Lindgren C, Baron J, Adami HO. Energy, nutrient intake and prostate cancer risk: a population-based case-control study in Sweden. Int J Cancer. 1996 Dec 11;68(6):716-22).

(Anon. 1999) (Anon. Dietary supplementation with n-3 polyunsaturated fatty acids and vitamin E after myocardial infarction: results of the GISSI-Prevenzione trial Gruppo Italiano per lo Studio della Sopravvivenza nell'Infarto miocardico. Lancet. 1999; 354:447–455. doi: 10.1016/S0140-6736(99)07072-5).

(Arain M.A., Abdul Qadeer A., 2010) (Arain M.A., Abdul Qadeer A. Systematic review on "Vitamin e and prevention of colorectal cancer" Pak. J. Pharm. Sci. 2010; 23:125–130).

(Arrowsmith et al, 1989) (Morbidity and mortality among low birth weight infants exposed to an intravenous vitamin E product, E-Ferol. JB Arrowsmith et al. Pediatrics. 1989 Feb; 83(2):244-9).

(Asgari et al. 2009) (Maryam M. Asgari, Sonia S. Maruti, Lawrence H. Kushi, Emily White. Antioxidant Supplementation and Risk of Incident Melanomas. Arch Dermatol. 2009;145(8):879-882).

(Averill MM, et al., 2008) (Averill MM, et al. Neither antioxidants nor genistein inhibit the progression of established atherosclerotic lesions in older apoE deficient mice. Atherosclerosis. 2009;203:82–88).

(Aw, 1999) (Aw TY., 1999. Molecular and cellular responses to oxidative stress and changes in oxidation–reduction imbalance in the intestine. Am. J. Clin. Nutr. 70, 557–565)

(Bakker et al., 2010) (G.C. Bakker, M.J. van Erk, L. Pellis, S. Wopereis, C.M. Rubingh, N.H. Cnubben, et al. An antiinflammatory dietary mix modulates inflammation and oxidative and metabolic stress in overweight men: a nutrigenomics approach. Am J Clin Nutr, 91 (2010), pp. 1044-1059).

(Baur et al., 2006) (Baur JA, Pearson KJ, Price NL, Jamieson HA, Lerin C, Kalra A, Prabhu VV, Allard JS, Lopez-Lluch G, Lewis K, Pistell PJ, Poosala S, Becker KG, Boss O, Gwinn D, Wang M, Ramaswamy S, Fishbein KW, Spencer RG, Lakatta EG, Le Couteur D, Shaw RJ, Navas P, Puigserver P, Ingram DK, de Cabo R, Sinclair DA. Resveratrol improves health and survival of mice on a high-calorie diet. Nature. 2006 Nov 16;444(7117):337-42).

(Bairati I, Meyer F, Jobin E, et al., 2006) (Bairati I, Meyer F, Jobin E, et al. Antioxidant vitamins supplementation and mortality: a randomized trial in head and neck cancer patients. Int J Cancer. 2006; 119:2221–4).

(Bareggi SR et al. 2009) (Bareggi SR et al. Effects of clioquinol on memory impairment and the neurochemical modifications induced by scrapie infection in golden hamsters. Brain Res. 2009 Jul 14; 1280:195-200). Department of Pharmacology, Chemotherapy and Medical Toxicology, School of Medicine, Milano, Italy).

(Baron et al, 2003) (John A. Baron, Bernard F. Cole, Leila Mott, Robert Haile, Maria Grau, Timothy R. Church, Gerald J. Beck, E. Robert Greenberg. Neoplastic and Antineoplastic Effects of Beta Carotene on Colorectal Adenoma Recurrence: Results of a Randomized Trial. JNCI Journal of the National Cancer Institute 2003 95(10):717-722).

(Beckman, Ames, 1998) (Beckman, K. & Ames, B. (1998) "The free radical theory of aging matures" Physiol Rev 78: 548-81).

(Begley, 2001) (Begley, The downside of antioxidants. Newsweek, Feb. 16, 2001).

(Benade et al, 1969) (Benade L, Howard T, Burk D. Synergistic killing of Ehrlich ascites carcinoma cells by ascorbate and 3-amino-1, 2, 4-triazole. Oncology 1969; 23: 33-43).

(Benfeitas R, Uhlen M, Nielsen J, Mardinoglu A., 2017) (Benfeitas R, Uhlen M, Nielsen J, Mardinoglu A. New Challenges to Study Heterogeneity in Cancer Redox Metabolism. Front Cell Dev Biol. 2017 Jul 11; 5:65).

(Berwick, Schantz, 1997) (Berwick M, Schantz S. Chemoprevention of aerodigestive cancer. Cancer Metastasis Rev. 1997 Sep-Dec;16(3-4) :529-47).

(Berube et al, 2008) (Sylvie Bérubé, Caroline Diorio and Jacques Brisson. Multivitamin-multimineral supplement use and mammographic breast density. American Journal of Clinical Nutrition, Vol. 87, No. 5, 1400-1404, May 2008)

(Birrell MA, McCluskie K, Wong S, Donnelly LE, Barnes PJ, Belvisi MG., 2005) (Birrell MA, McCluskie K, Wong S, Donnelly LE, Barnes PJ, Belvisi MG. Resveratrol, an extract of red wine, inhibits lipopolysaccharide induced airway neutrophilia and inflammatory mediators through an NF-kappaB-independent mechanism. FASEB J. 2005 May;19(7):840-1) (Rahman I, Biswas SK, Kirkham PA., 2006)

(Bjelakovic et al. 2004) (Bjelakovic G, Nikolova D, Simonetti RG, Gluud C. Antioxidant supplements for preventing gastrointestinal cancers. Cochrane Database Syst Rev. doi: 10.1002/14651858.CD004183.pub2. 2004;(4):CD004183).

(Bjelakovic G., Nikolova D., Simonetti R.G., et al., 2004) (Bjelakovic G., Nikolova D., Simonetti R.G., Gluud C. Antioxidant supplements for prevention of gastrointestinal cancers: A systematic review and meta-analysis. Lancet. 2004; 364:1219–1228).

(Bjelakovic et al, 2006) (Bjelakovic G, Nagorni A, Nikolova D, Simonetti RG, Bjelakovic M, Gluud C. Meta-analysis: antioxidant supplements for primary and secondary prevention of colorectal adenoma. Aliment Pharmacol Ther. 2006; 24:281-291).

(Bjlakovic et al., pp. 842-57, 2007) (Bjelakovic G, Nikolova D, Gluud LL, Simonetti RG, and Gluud C. "Mortality in Randomized Trials of Antioxidant Supplements for Primary and Secondary Prevention; Systematic Review and Meta-analysis." JAMA 2007; 297:842-857. Vol. 297 No. 8, February 28, 2007).

(Bjlakovic et al., pp. 765-6, 2008) (Bjelakovic G, Nikolova D, Gluud LL, Simonetti RG, and Gluud C. "Mortality in Randomized Trials of Antioxidant Supplements for Primary and Secondary Prevention; Systematic Review and Meta-analysis." JAMA 2007; 299: 765-6. Vol. 299 No. 8, February 20, 2008).

(Bjelakovic, Nikolova, Simonette and Gludd, 2008 Sept) (Bjelakovic G, Nikolova D, Simonetti RG, Gluud C. Systematic review: primary and secondary prevention of gastrointestinal cancers with antioxidant supplements. Aliment Pharmacol Ther. 2008 Sep 15;28(6):689-703).

(Bjelakovic, Nikolova, Simonette and Gludd, 2012 Mar. 14) ((Bjelakovic, Nikolova, Simonette and Gludd, Antioxidant supplements for prevention of mortality in healthy participants and patients with various diseases. ... 14 Mar 2012 | DOI: 10.1002/14651858.CD007176.pub2).

(Bleys et al, 2007) (Selenium and Diabetes: More Bad News for Supplements. Joachim Bleys, Ana Navas-Acien, and Eliseo Gualla. Ann Intern Med 2007; 147: 271-272).

(Blot WJ, Li JY, Taylor PR, et al., 1993) (Blot WJ, Li JY, Taylor PR, Guo W, Dawsey S, Wang GQ, et al. Nutrition intervention trials in Linxian, China: supplementation with specific vitamin/mineral combinations, cancer incidence, and disease-specific mortality in the general population. J Natl Cancer Inst. 1993; 85:1483-92).

(Bobe G., Weinstein S.J., Albanes D., et al., 2008) (Bobe G., Weinstein S.J., Albanes D., Hirvonen T., Ashby J., Taylor P.R., Virtamo J., Stolzenberg-Solomon R.Z. Flavonoid intake and risk of pancreatic cancer in male smokers (finland) Cancer Epidemiol. Biomark. Prev. 2008; 17:553–562).

(Boffeta P, Wichmann J, Ferrari P, et al., 2010) (Boffeta P, Wichmann J, Ferrari P, et al. Fruit and vegetable intake and overall cancer risk in the European Prospective Investigation into Cancer and Nutrition (EPIC). J Natl Cancer Inst. 2010; 102:529–37).

(Boehm et al, 2009) (Boehm et al. Green tea (Camellia sinensis) for the prevention of cancer. Chochrane Database Syst Rev. 2009 Jul 8;(3):CD005004).

(Boos, Stopper, 2000) (Boos, G., Stopper, H., Genotoxicity of several clinically used topoisomerase II inhibitors. Toxicol. Lett. 2000, 116, 7–16).

(Borek C., 1997) (Borek C. Antioxidants and cancer. Sci Med (Phila). 1997; 4:51–62)

(Borek C, Pardo F., 2002) (Borek C, Pardo F. Vitamin E and apoptosis: a dual role. In: Pasquier C, ed. Biennial Meeting of the Society for Free Radicals Research International: Paris, France July 16–20, 2002. Bologna, Italy: Medimond Inc. Monduzzi (Editore); 2002. pp. 327–31).

(Borek, 2004) (Borek C. Antioxidants and radiation therapy. J Nutr. 2004; 134:3207S–9S).

(Borek, Carmia, 2017) (Borek, Carmia, Dietary Antioxidants and Human Cancer. Journal of Restorative Medicine, Volume 6, Number 1, 7 December 2017, pp. 53-61(9).

(Bouayed J, Bohn T., 2010) (Bouayed J, Bohn T. Exogenous antioxidants – double-edged sword in cellular redox state: health beneficial effects at physiologic doses versus deleterious effects at high doses. Oxid Med Cell Longev. 2010; 3:228–37).

(Bove et al, 1985) (Vasculopathic hepatotoxicity associated with E-Ferol syndrome in low-birth-weight infants. K. E. Bove, N. Kosmetatos, K. E. Wedig, D. J. Frank, S. Whitlatch, V. Saldivar, J. Haas, C. Bodenstein and W. F. Balistreri. JAMA. Vol. 254. No. 17. November 1, 1985).

(Bowman T.S., Bassuk S.S., Gaziano M., 2007) (Bowman T.S., Bassuk S.S., Gaziano M. Interventional Trials of Antioxidants. In: Hotzman J.L., editor. Atherosclerosis and Oxidant Stress: A New Perspective. Springer; New York, NY, USA: 2007. pp. 25–50).

(Brambilla et al, 2008) (Brambilla D, et al. The role of antioxidant supplement in immune system, neoplastic, and neurodegenerative disorders: a point of view for an assessment of the risk/benefit profile. Nutr J. 2008 Sep 30; 7:29).

(Breinholt et al. 2003) (Breinholt VM, et al. "Food and Chemical Toxicology"; Effects of Dietary Antioxidants and 2-Amino-3-Methylimidazo[4,5-f]-Quinoline on Preneoplastic Lesions and on Oxidative Damage, Hormonal Status and Detoxification Capacity in the Rat; October 2003. Vol 41. Issue 10, Pp 1315-1323).

(Brown et al, 2001) (Brown BG, Zhao XQ, Chait A, Fisher LD, Cheung MC, Morse JS, Dowdy AA, Marino EK, Bolson EL, Alaupovic P, Frohlich J, and Albers JJ. Simvastatin and niacin, antioxidant vitamins, or the combination for the prevention of coronary disease. N Engl J Med 345: 1583–1592, 2001).

(Brown GC, Borutaite V., 2012) (Brown GC, Borutaite V., There is no evidence that mitochondria are the main source of reactive oxygen species in mammalian cells. Mitochondrion. 2012 Jan;12(1):1-4).

(Buffenstein, 2008, CPB) (Buffenstein R. Negligible senescence in the longest living rodent, the naked mole-rat: insights from a successfully aging species. J Comp Physiol B. 2008;178:439–45)

(Bruunsgaard et al., 2003) (H. Bruunsgaard, H.E. Poulsen, B.K. Pedersen, K. Nyyssönen, J. Kaikkonen, J.T. Salonen. Long-term combined supplementations with α-tocopherol and vitamin C have no detectable anti-inflammatory effects in healthy men. J Nutr, 133 (2003), pp. 1170-1173).

(Cadenas E., 2004) (Cadenas E. Mitochondrial free radical production and cell signaling. Mol. Aspects Med. 2004;25:17–26).

(Calzada et al, 1997) (Calzada C, Bruckdorfer KR, Rice-Evans CA. The influence of antioxidant nutrients on platelet function in healthy volunteers. Atherosclerosis 1997 Jan 3;128(1):97-105)

(Caraballoso et al, 2003) (Caraballoso M, Sacristan M, Serra C, Bonfill X. Drugs for preventing lung cancer in healthy people. Cochrane Database Syst Rev. doi: 10.1002/14651858.CD002141. 2003;(2):CD002141).

(Chandra et al, 2003) (Chandra J, Hackbarth J, Le S, et al. Involvement of reactive oxygen species in adaphostin-induced cytotoxicity in human leukemia cells. Blood. 2003;102: 4512-4519).

(Chapman et al, 1991) (Chapman, J.D., Stobbe, C.C., Arnfield, M.R., Santus, R., Lee, L. and McPhee, M.S. Oxygen dependency of tumor cell killing in vitro by light-activated Photofrin II. Radiat Res 1991; 126: 73-79).

(Chapmann, 1979) (William Chapmann, "A Japanese Tragedy, Dirodohydroxyquinoline," Washington Post 18 March 1979).

(Cheung M.C., Zhao X.Q., Chait A., 2001) (Cheung M.C., Zhao X.Q., Chait A., Albers J.J., Brown B.G. Antioxidant supplements block the response of hdl to simvastatin-niacin therapy in patients with coronary artery disease and low hdl. Arterioscler. Thromb. Vasc. Biol. 2001; 21:1320–1326).

(Chiu-Lan Hsieh et al, 2010) (Chiu-Lan Hsieh, Chiung-chi Peng, Yu-Ming Cheng, et al. Quercetin and Ferulic Acid Aggravate Renal Carcinoma in Long-Term Diabetic Victims. J. Agric. Food Chem., 2010, 58 (16), pp 9273–9280).

(Choi, Singh, 2005) (S. Choi and S.V. Singh. Bax and Bak are required for apoptosis induction by sulforaphane, a cruciferous vegetable-derived cancer chemopreventive agent. Cancer Res. 2005 Mar 1; 65(5):2035-43).

(Cohen and Krasnow, 1987) (Cohen M H, Krasnow S H. Cure of advanced Lewis lung carcinoma (LL): a new treatment strategy. Proceedings of AACR 1987; 28: 416).

(Collins et al, 2002) (Collins, R., J. Armitage, S. Parish, P. Sleight & R. Peto. 2002. MRC/BHF Heart Protection Study of antioxidant vitamin supplementation in 20,536 high-risk individuals: A randomized placebo-controlled trial. Lancet. 360(9326):23-33).

(Cook et al, 2004) (Cook JA, Gius D, Wink DA, Krishna MC, Russo A, Mitchell JB: Oxidative stress, redox, and the tumor microenvironment. Semin Radiat Oncol 14: 259-266, 2004)

(Czernichow et al, 2006) (Antioxidant supplementation does not affect fasting plasma glucose in the Supplementation with Antioxidant Vitamins and Minerals (SU.VI.MAX) study in France: association with dietary intake and plasma concentrations. S. Czernichow, A. Couthouis, S. Bertrais, A.-C. Vergnaud, L. Dauchet, P. Galan, and S. Hercberg. Am. J. Clinical Nutrition, August 1, 2006; 84(2): 395 - 399)

(Czernichow et al, 2009) (Czernichow, S. et al. Effects of long-term antioxidant supplementation and association of serum antioxidant concentrations with risk of metabolic syndrome in adults. Am. J. Clinical Nutrition, Vol. 90, No. 2, 329-335, August 2009)

(Damiani E., Astolfi P., Carloni P., Stipa P., Greci L., 2008) (Damiani E., Astolfi P., Carloni P., Stipa P., Greci L. Antioxidants: How They Work. In: Valacchi G., Davis P.A., editors. Oxidants in Biology. Springer Science + Buisness Media; New York, NY, USA: 2008. pp. 251–266)

(Davies, 1999) (Davies KJ. The broad spectrum of responses to oxidants in proliferating cells: a new paradigm for oxidative stress. IUBMB Life. 1999 Jul; 48(1):41-7).

(de Gast et al, 2000) (de Gast GC, Klümpen HJ, Vyth-Dreese FA; et al. Phase I trial of combined immunotherapy with subcutaneous granulocyte macrophage colony-stimulating factor, low-dose interleukin 2, and interferon alpha in progressive metastatic melanoma and renal cell carcinoma. Clin Cancer Res. 2000;6(4):1267-1272).

(Dewella A, Tsaob P, Rigdon J, et al., 2018) (Dewella A, Tsaob P, Rigdon J, et al. Antioxidants from diet or supplements do not alter inflammatory markers in adults with cardiovascular disease risk. A pilot randomized controlled trial. Nutrition Research. Volume 50, February 2018, Pages 63-72).

(Divi, Doerge, 1996) (Divi RL, Doerge DR. Inhibition of thyroid peroxidase by dietary flavonoids. Chem Res Toxicol. 1996 Jan-Feb;9(1):16-23).

(Donghui et al., 2009) (Donghui et al. Gastroenterology, August 2009;137(2):482–488).

(Donnelly et al, Mutagenesis. 1999) (Donnelly ET, McClure N, Lewis SE. The effect of ascorbate and alpha-tocopherol supplementation in vitro on DNA integrity and hydrogen peroxide-induced DNA damage in human spermatozoa. Mutagenesis. 1999 Sep;14(5):505-12).

(Doroshenko et al, 2004) (Doroshenko N, Doroshenko P: The glutathione reductase inhibitor carmustine induces an influx of Ca2+ in PC12 cells. Eur J Pharmacol 497: 17-24, 2004)

(Dotan, Lichtenberg and Pinchuk, 2009) (Dotan Y, Lichtenberg D, Pinchuk I. No evidence supports vitamin E indiscriminate supplementation. Biofactors. 2009 Nov-Dec;35(6):469-73)

(Duffield-Lillico, 2003) (Duffueld-Lillico AJ, et al, Selenium supplementation and secondary prevention of nonmelanoma skin cancer in a randomized trial. J Natl Cancer Inst. 2003 Oct 1;95(19):1477-81).

(Dündar Y., Aslan R., 2000) (Dündar Y., Aslan R. Antioxidative stress. Eastern J. Med. 2000; 5:45–47).

(Duranteau et al, 1998) (Duranteau J, Chandel NS, Kulisz A, Shao Z, Schumacker PT, 1998. Intracellular signaling by reactive oxygen species during hypoxia in cardiomyocytes. J. Biol. Chem. 273, 11619–11624).

(Dussan et al, 2008) (Dussan C, Zubor P, Fernandez M, Yabar A, Szunyogh N, Visnovsky J. Spontaneous regression of a breast carcinoma: a case report. Gynecol Obstet Invest. 2008;65(3):206-211).

(Egert et al., 2009) (S. Egert, A. Bosy-Westphal, J. Seiberl, C. Kürbitz, U. Settler, S. Plachta-Danielzik, et al. Quercetin reduces systolic blood pressure and plasma oxidised low-density lipoprotein concentrations in overweight subjects with a high-cardiovascular disease risk phenotype: a double-blinded, placebo-controlled cross-over study. Br J Nutr, 102 (2009), pp. 1065-1074).

(Enns J, et al., 2014) (Enns J, et al. The impact of omega-3 polyunsaturated fatty acid supplementation on the incidence of cardiovascular events and complications in peripheral arterial disease: a systematic review and meta-analysis. BMC Cardiovasc Disord. 2014; 14:70).

(Evans, Lawrenson. 2017) (Jennifer R Evans, John G Lawrenson. Antioxidant vitamin and mineral supplements for preventing age-related macular degeneration. Cochrane Eyes and Vision Group. 30 July 2017. DOI: 10.1002/14651858.CD000253.pub4).

(Ezzedine et al, 2010) (Ezzedine et al. Incidence of skin cancers during 5-year follow-up after stopping antioxidant vitamins and mineral supplementation. Eur J Cancer. 2010Dec;46(18):3316-22).

(Falcone et al, 2010) (E Liana Falcone, Alexandra Mangili, Alice M Tang, Clara Y Jones, Margo N Coods, Joseph F Polak and Christine A. Wanke. Micronutrient concentrations and subclinical atherosclerosis in adults with HIV. Am J Clin Nutr 91: 1213-1219, 2010. Vol. 91, No. 5, 1213-1219, May 2010).

(Fang, Nakamura, Iyer, 2007) (Fang J, Nakamura H, Iyer AK. Tumor-targeted induction of oxystress for cancer therapy. J Drug Target. 2007 Aug-Sep;15(7-8):475-86)

(Figueiredo et al, 2009) (Figueiredo JC et al. Folic acid and risk of prostate cancer: results from a randomized clinical trial. J Natl Cancer Inst. 2009 Mar 18;101(6):432-5).

(Fisher-Wellman K., Bell H.K., Bloomer R.J., 2009) (Fisher-Wellman K., Bell H.K., Bloomer R.J. Oxidative stress and antioxidant defense mechanisms linked to exercise during cardiopulmonary and metabolic disorders. Oxid. Med. Cell. Longev. 2009; 2:43–51).

(Forbes et al, 2003) (Forbes K, Gillette K, Sehgal I. Lycopene increases urokinase receptor and fails to inhibit growth or connexin expression in a metastatically passaged prostate cancer cell line: a brief communication. Exp Biol Med (Maywood). 2003 Sep;228(8):967-71)

(Fortmann SP, Burda BU, Senger CA, et al., 2013) (Fortmann SP, Burda BU, Senger CA, et al. Vitamin and mineral supplements in the primary prevention of cardiovascular disease and cancer: an updated systematic evidence review for the U.S. Preventive Services Task Force. Ann Intern Med. 2013; 159:824–34).

(Frankel et al, 2014) (Frankel PH, Parker RS, Madsen FC, Whanger PD. Baseline selenium and prostate cancer risk: comments and open questions. J Natl Cancer Inst. Natl Cancer Inst. 2014 Feb 22. [Epub ahead of print]).

(Gaziano JM, Glynn RJ, Christen WG, et al., 2009) (Gaziano JM, Glynn RJ, Christen WG, et al. Vitamins E and C in the prevention of prostate and total cancer in men: The Physicians' Health Study II randomized controlled trial. J Am Med Assoc. 2009; 301:52–62).

(Gibbons et al, 2003) (Gibbons RJ, Abrams J, Chatterjee K, et al; American College of Cardiology/American Heart Association Task Force on Practice Guidelines. Committee on the Management of Patients with Chronic Stable Angina. ACC/AHA 2002 guideline update for the management of patients with chronic stable angina—summary article: a report of the American College of Cardiology/American Heart Association Task Force on Practice Guidelines (Committee on the Management of Patients with Chronic Stable Angina). Circulation. 2003; 107: 149–158)

(Gershman et al, 1954) (Gershman R, Gilbert DL, Nye SW, Dwyer P, Fenn WO. Oxygen poisoning and X-irradiation: a mechanism in common. Science. 1954;119:623–6)

(Giovannucci, et al., 1998) (Giovannucci, E., M.J. Stampfer, G.A. Colditz, D.J. Hunter, C. Fuchs, B.A. Rosner, F.E. Speizer & W.C. Willett. 1998. Multivitamin use, folate, and colon cancer in women in the Nurses' Health Study. Ann. Intern. Med. 129(7):517-524).

(GISSI, 1999) (Dietary supplement with n-3 polyunsaturated acids and vitamin E after myocardial infarction: results of the GISSI-Prevention trial. Gruppo, Italiano per lo Studio Sopravvivenza nell'Infarto miocardico. 1999. Lancet. 354: 447-455).

(Gleave et al, 1998) (Gleave ME, Elhilali M, Fradet Y; et al, Canadian Urologic Oncology Group. Interferon gamma-1b compared with placebo in metastatic renal-cell carcinoma. N Engl J Med. 1998;338(18):1265-1271).

(Gomez-Cabrera M.C., Domenech E., Vina J., 2008) (Gomez-Cabrera M.C., Domenech E., Vina J. Moderate exercise is an antioxidant: Upregulation of antioxidant genes by training. Free Radic. Biol. Med. 2008;44:126–131).

(Goodman GE, Thornquist MD, Balmes J, et al., 2004) (Goodman GE, Thornquist MD, Balmes J, et al. The Beta-Carotene and Retinol Efficacy Trial: incidence of lung cancer and cardiovascular disease mortality during 6-year follow-up after stopping beta-carotene and retinol supplements. J Natl Cancer Inst. 2004; 96:1743–50).

(Greenlee et al, 2009) (Greenlee H, Gammon MD, Abrahamson PE, et al. Prevalence and predictors of antioxidant supplement use during breast cancer treatment: The Long Island Breast Cancer Study Project. Cancer. 2009 Jul 15;115(14):3271-82).

(Greenlee, Hershman et al, 2009) (Greenlee H, Hershman DL, Jacobson JS. Use of antioxidant supplements during breast cancer treatment: a comprehensive review. Breast Cancer Res Treat. 2009 Jun;115(3):437-52).

(Griffiths K, Aggarwal BB, Singh RB, et al., 2016) (Griffiths K, Aggarwal BB, Singh RB, et al. Food antioxidants and their anti-inflammatory properties: a potential role in cardiovascular diseases and cancer prevention. Diseases. 2016; 4: E28).

(Gu G.J., et al., 2008) (Gu G.J., Li Y.P., Peng Z.Y., Xu J.J., Kang Z.M., Xu W.G., Tao H.Y., Ostrowski R.P., Zhang J.H., Sun X.J. Mechanism of ischemic tolerance induced by hyperbaric oxygen preconditioning involves upregulation of hypoxia-inducible factor-1alpha and erythropoietin in rats. J. Appl. Physiol. 2008;104:1185).

(Gupta et al, 2011) (Subash C. Gupta, David Hevia, Sridevi Patchva, Byoungduck Park, Wonil Koh, and Bharat B. Aggarwal. Upsides and Downsides of Reactive Oxygen Species for Cancer: The Roles of Reactive Oxygen Species in Tumorigenesis, Prevention, and Therapy. ANTIOXIDANTS & REDOX SIGNALING Volume 16, Number 11, 2012. Pp. 1295-1322. DOI: 10.1089/ars.2011.4414).

(Gutierrez J., Ballinger S.W., Darley-Usmar V.M., Landar A., 2006) (Gutierrez J., Ballinger S.W., Darley-Usmar V.M., Landar A. Free radicals, mitochondria, and oxidized lipids: The emerging role in signal transduction in vascular cells. Circ. Res. 2006;99:924–932).

(Gutteridge J.M., Halliwell B., 2010) (Gutteridge J.M., Halliwell B. Antioxidants: Molecules, medicines, and myths. Biochem. Biophys. Res Commun. 2010; 393:561–564).

(Hak et al. 2003) (Hak AE, Stampfer MJ, Campos H, Sesso HD, Gaziano JM, Willett W, Ma J: Plasma carotenoids and tocopherols and risk of myocardial infarction in a low-risk population of US male physicians. Circulation 108:802: E9012 –E9013,2003)

(Halliwell B., Gutteridge J.M., 1995) (Halliwell B., Gutteridge J.M. The definition and measurement of antioxidants in biological systems. Free Radic. Biol. Med. 1995;18:125–126).

(Halliwell and Gutteridge, 1998) (Halliwell, B. & Gutteridge, J.M.C. in Free Radicals in Biology and Medicine, 3rd edition. (Oxford University Press, UK, 1999) (Hampton, M.B., Fadeel, B., Orrenius, S. Redox regulation of the caspases during apoptosis. Ann. NY Acad. Sci. 854, 328–335 (1998).

(Halliwell, Guteridge, 1989) (Halliwell B, Gutteridge J M C, eds. Free radicals in biology and medicine, 2nd edn. Oxford: Clarendon Press, 1989; 124).

(Halliwell and Gutteridge, 1999) (Halliwell, B. & Gutteridge, J.M.C. in Free Radicals in Biology and Medicine, 3rd edition. (Oxford University Press, UK, 1999).

(Halliwell, 2000) (A super way to kill cancer cells? Barry Halliwell. Nature Medicine 6, 1105 - 1106 (2000).

(Halliwell B., 2007) (Halliwell B. Biochemistry of oxidative stress. Biochem. Soc. Trans. 2007;35:1147–1150).

(Halliwell B., 2008) (Halliwell B. Are polyphenols antioxidants or pro-oxidants? What do we learn from cell culture and in vivo studies? Arch. Biochem. Biophys. 2008;476:107–112).

(Halliwell B., 2009) (Halliwell B. The wanderings of a free radical. Free Radic. Biol. Med. 2009;46:531–542).

(Halliwell B., Lee C.Y., 2010) (Halliwell B., Lee C.Y. Using isoprostanes as biomarkers of oxidative stress: Some rarely considered issues. Antioxid. Redox Signal. 2010;13:145–156).

(Halliwell B., 2012) (Halliwell B. Free radicals and antioxidants: updating a personal view. Nutr Rev. 2012; 70:257–65).

(Hampton et al, 1995) (Hampton, M.B., Fadeel, B., Orrenius, S. Redox regulation of the caspases during apoptosis. Ann. NY Acad. Sci. 854, 328–335 (1998) (Jacobson, M.D. & Raff, M.C. Programmed cell death and Bcl-2 protection in very low oxygen. Nature 374, 814–816 (1995).

(Hampton, Orrenius, 1997) (MB Hampton, S Orrenius. Dual regulation of caspase activity by hydrogen peroxide: implications for apoptosis. FEBS Lett (1997) 414: 552-6).

(Hampton et al, 1998) (Hampton, M.B., Fadeel, B., Orrenius, S. Redox regulation of the caspases during apoptosis. Ann. NY Acad. Sci. 1998. 854, 328–335).

(Hara et al, 2011) (Hara A, et al. Use of vitamin supplements and risk of total cancer and cardiovascular disease among the Japanese general population: a population-based survey. BMC Public Health. 2011 Jul 8;11:540).

(Harman, 1956) (Harman D. Aging: a theory based on free radical and radiation chemistry. J Gerontol. 1956;11:298–300)

(Harris, 2017) (Harris R. Rigor Mortis: How slopy science creates worthless cures, Crushes hope and wastes billions, New York, NY: Basic Books, 2017)

(Hasnain B.I., Mooradian A.D., 2004) (Hasnain B.I., Mooradian A.D. Recent trials of antioxidant therapy: What should we be telling our patients? Cleve. Clin. J. Med. 2004;71:327–334).

(Hayden et al, 2007) (K. Hayden, K. Welsh-Bohmer, H. Wengreen, P. Zandi, C. Lyketsos, J. Breitner. Risk of Mortality with Vitamin E Supplements: The Cache County Study. Am J Med. 2007 Feb;120(2):180-4)

Heart Protection Study Collaborative Group. Mrc/bhf, 2002) (Heart Protection Study Collaborative Group. Mrc/bhf heart protection study of antioxidant vitamin supplementation in 20,536 high-risk individuals: A randomised placebo-controlled trial. Lancet. 2002;360:23–33).

(Helferich et al, 2008) (Helferich WG, et al. Phytoestrogens and breast cancer: a complex story. Inflammopharmacology. 2008 Oct;16(5):219-26)

(Heinonen et al, 1994) (Heinonen, O.P., J.K. Huttunen, D. Albanes & ATBC cancer prevention study group. 1994. The effect of vitamin E and beta carotene on the incidence of lung cancer and other cancers in male smokers. N. Engl. J. Med. 330:1029-1035).

(Helzlsouer KJ, Huang HY, Alberg AJ, et al., 2000) (Helzlsouer KJ, Huang HY, Alberg AJ, et al. Association between alpha-tocopherol, gamma-tocopherol, selenium, and subsequent prostate cancer. J Natl Cancer Inst. 2000; 92:2018–23).

(Hemila and Kaprio, 2009 Apr) (Hemilä H, Kaprio J. Modification of the effect of vitamin E supplementation on the mortality of male smokers by age and dietary vitamin C. Am J Epidemiol. 2009 Apr 15;169(8):946-53)

(Hennekens CH, Buring JE, Manson JE, et al., 1996) (Hennekens CH, Buring JE, Manson JE, et al. Lack of effect of long-term supplementation with beta carotene on the incidence of malignant neoplasms and cardiovascular disease. N Engl J Med. 1996; 334:1145–9).

(Hercberg S, Galan P, Preziosi P, et al., 2004) (Hercberg S, Galan P, Preziosi P, et al. The SU.VI.MAX study: a randomized, placebo-controlled trial of the health effects of antioxidant vitamins and minerals. Arch Intern Med. 2004; 164:2335–42).

(Hercberg et al, 2007) (Serge Hercberg et al. Antioxidant Supplementation Increases the Risk of Skin Cancers in Women but Not in Men. American Society for Nutrition J. Nutr. 137:2098-2105, September 2007).

(Hirsch, Iliopoulos, Tsichlis and Struhl, 2009) (Hirsch, Iliopoulos, Tsichlis and Struhl. Metformin Selectively Targets Cancer Stem Cells and Acts Together with Chemotherapy to Block Tumor Growth and Prolong Remission. Cancer Res, Oct 1, 2009;69(19):7507-11).

(Ho et al, 2007) (Ho et al, 2007) (J. C. Ho et al. Disturbance of systemic antioxidant profile in nonsmall cell lung carcinoma. Eur Respir J 2007; 29:273-278).

(Hodis HN., 2002) (Hodis HN. Alpha-tocopherol supplementation in healthy individuals reduces low-density lipoprotein oxidation but not atherosclerosis: The Vitamin E Atherosclerosis Prevention Study (VEAPS) Circulation. 2002;106:1453–1459).

(Honarbakhsh, Schachter, 2009) (Honarbakhsh S, Schachter M. Vitamins and cardiovascular disease. Br J Nutr. 2009 Apr;101(8:1113-31)

(Hooper L, et al., 2006) (Hooper L, et al. Risks and benefits of omega 3 fats for mortality, cardiovascular disease, and cancer: systematic review. BMJ. 2006;332:752–760).

(Howes, Steele, 1971) (Howes, R. M. and Steele, R. H., Microsomal chemiluminescence induced by NADPH and its relation to lipid peroxidation, Res. Commun. Chem. Path. Pharmacol., July-Sept. 1971, 2; 4 & 5:619-626).

(Howes, Steele, 1972) (Howes, R.M. and Steele, R.H., Microsomal chemiluminescence induced by NADPH and its relation to aryl-hydroxylations, Res Commun. Chem. Path. Pharmacol., March 1972, 3; 2:349-357).

(Howes et al, 1976) (Howes, R. M., Allen, R.C., Su, C.T. and Hoopes, J.E., Altered polymorphonuclear leukocyte bioenergetics in patients with thermal injury, the Surgical Forum, 1976, 27:558-560).

(Howes, Steele, 1976) (Howes, R.M., Steele, R.H. and Hoopes, J.E., Peroxide induced Chemiluminescence in an in vitro proline hydroxylation system, 1976, 8; 1:77-84).

(Howes et al, 1977) (Howes, R.M., Steele, R.H. and Hoopes, J.E., The role of Electronic excitation states in collagen biosynthesis, Persp. In Biol. And Med., Summer 1977, 20; 4:539-544).

(Howes, UTOPIA, 2004) (Howes, R. M. U.T.O.P.I.A. - Unified Theory of Oxygen Participation in Aerobiosis. © 2004. Free Radical Publishing Co. Kentwood, LA, available at www.iwillfindthecure.org.).

(Howes, 2005) (Howes, R.M. Tumoricidal Activity of an Injectable Singlet Oxygen System Generated from Physiological Agents: The Howes Singlet Oxygen Cancer Therapy System). In The Medical and Scientific Significance of Oxygen Free Radical Metabolism. © 2005. Free Radical Publishing Co. Kentwood, LA. pp. 893-912).

(Howes, Farber, 2005) (Howes, R.M. and Farber, G. Tumoricidal Activity of the Howes Singlet Oxygen Delivery System in Human Basal Cell Carcinoma. In The Medical and Scientific Significance of Oxygen Free Radical Metabolism. © 2005. Free Radical Publishing Co. Kentwood, LA. pp. 883-892).

(Howes, 2006) (Howes R.M. The Free Radical Fantasy: A Panoply of Paradoxes. Ann. N. Y. Acad. Sci. 2006; 1067:22-26).

(Howes, Philica. Feb 26, 2007) (Howes M.D., PhD., R. (2007). Cancer, Apoptosis and Reactive Oxygen Species: A New Paradigm. PHILICA.COM Article number 86. Published on 26th February 2007).

(Howes, Philica. April 5, 2007) (Howes M.D., PhD., R. (2007). Antioxidant Vitamins A, C & E; Death in Small Doses and Legal Liability? PHILICA.COM Article number 89. Published April 5, 2007).

(Howes, Am J Cos Surg. 2009) (Antioxidant Vitamins: A Review of Policy Statements and Recommendations. R.M. Howes. The American Journal of Cosmetic Surgery. Vol. 26, No. 2, pg. 63-78, 2009).

(Howes, Philica. Feb 7, 2009) (Howes M.D., PhD., R. (2009). Dangers of Antioxidants in Cancer Patients: A Review. PHILICA.COM Article number 153. Published 7th February 2009).

(Howes, 2009) (Howes, R. M. Reactive Oxygen Species Insufficiency (ROSI) as the Basis for Disease Allowance and Coexistence. © 2009. Free Radical Publishing Co. Kentwood, LA, available at www.iwillfindthecure.org.).

(Howes R: Hydrogen Peroxide: 2010) (R. Howes: Hydrogen Peroxide: A review of a scientifically verifiable omnipresent ubiquitous essentiality of obligate, aerobic, carbon-based life forms. The Internet Journal of Plastic Surgery. 2010 Volume 7 Number 1)
(Howes, 2010) (Howes R. Death in Small Doses? Trafford Publishing. Indianapolis, USA. 2010).

(Howes R: Cancer Therapy, 2010) (Howes R: Cancer Therapy: A Review with Scientific Validation for the Role of Electronically Modified Oxygen Derivatives in Oncologic Treatment Modalities. The Internet Journal of Alternative Medicine. 2010 Volume 8 Number 1).

(Howes RM. Death in Small Doses: Book One, 2010) (Howes RM. Death in Small Doses: Book One, Trafford Publishing. Indianapolis, USA. 2010)

(Howes RM. Death in Small Doses: Book Two, 2010) (Howes RM. Death in Small Doses: Book Two, Trafford Publishing. Indianapolis, USA. 2010)

(Howes, Overkill, 2011) (Antioxidant Overkill, Prof. R.M. Howes MD, PhD. CreateSpace and Free Radical Publishing, © 2011).

(Howes, Dangers, 2011) (Dangers of Excessive Antioxidants in Cancer Patients, Prof. R. M. Howes MD, PhD. CreateSpace and Free Radical Publishing © 2011).

(Howes, Heart disease, 2011) (Heart Disease and Antioxidant Failures, Prof. R.M. Howes MD, PhD. CreateSpace and Free Radical Publishing, © 2011)

(Howes, AOX failures, 2011) (Antioxidant Failures and Dangers, Prof. R.M. Howes MD, PhD. CreateSpace and Free Radical Publishing, © 2011).

(Howes, antiaging, 2011) (Anti-Aging Anti-Oxidant Scams, Prof. R.M. Howes MD, PhD. CreateSpace and Free Radical Publishing, © 2011).

(Howes, Sports, 2011) (Sports, Athletes, Exercise Facts and Antioxidant Myths, Prof. R.M. Howes MD, PhD. CreateSpace and Free Radical Publishing, © 2011).

(Howes, Alzheimer's, 2012) (Alzheimer's Disease: Forget Antioxidants and Supplements, Prof. R.M. Howes MD, PhD. CreateSpace and Free Radical Publishing, © 2012).

(Howes, Sex, 2012) (Sex, Performance, Reproduction, Naked Radicals and Antioxidants, Prof. R.M. Howes MD, PhD. CreateSpace and Free Radical Publishing, © 2012).

(Howes, Linked, 2012) (Antioxidant Links to Deadly Unintended Consequences. Prof. R.M. Howes MD, PhD, CreateSpace and Free Radical Publishing Co, USA, © 2012).

(Howes, UTOPIA, 2014) (U.T.O.P.I.A.: Unified Theory of Oxygen Participation in Aerobiosis, Prof. R.M. Howes (MD, PhD. CreateSpace and Free Radical Publishing, © 2014, revised).

(Howes, H2O2, 2014) (Hydrogen Peroxide: A Health, Homeostatic and Protective Essentiality, Prof. R.M. Howes MD, PhD. CreateSpace and Free Radical Publishing, © 2014).

(Howes, Oxypocalypse, 2014) (Reactive Oxygen Species vs. Antioxidants: The Oxypocalypse or The War That Never Was, Prof. R.M. Howes MD, PhD. CreateSpace and Free Radical Publishing, © 2014).

(Howes, Diabetes, 2014) (Diabetes and Oxygen Free Radical Sophistry, Prof. R.M. Howes MD, PhD. CreateSpace and Free Radical Publishing, © 2014, revised).

(Howes, Fish oil, 2014) (FISH OIL (Omega3 fatty acids): Facts, Fantasies & Failures. Prof. R.M. Howes MD, PhD. CreateSpace and Free Radical Publishing, © 2014).

(Howes, Vit D, 2014) (Vitamin D: Benefits & False claims. Prof. R.M. Howes MD, PhD. CreateSpace and Free Radical Publishing, © 2014).

(Howes, Chocolate, wine, 2015) (Chocolate & Red Wine Antioxidants (Polyphenols, Flavonoids & Resveratrol): Facts vs. (Falsehoods. Prof. R.M. Howes MD, PhD. CreateSpace and Free Radical Publishing, © 2015).

(Howes, Chocolate, wine, 2015) (Chocolate & Red Wine Antioxidants (Polyphenols, Flavonoids & Resveratrol): Facts vs. (Falsehoods. Prof. R.M. Howes MD, PhD. CreateSpace and Free Radical Publishing, © 2015).

(Howes, Exercise, 2015) (Exercise and Reactive Oxygen Species. Likely the only health miracle out there. Prof. R.M. Howes MD, PhD. CreateSpace and Free Radical Publishing, © 2015).

(Howes, Cancer/longevity, 2015) (Cancer and Longevity Answers: Naked mole rats, Exercise & EMODs (ROS). Prof. R.M. Howes MD, PhD. CreateSpace and Free Radical Publishing, © 2015).

(Howes, HBOT, 2015) (Hyperbaric Oxygen, Hypoxia, Hyperoxia & EMODS (ROS): Separating Fact from Factitious. Prof. R.M. Howes MD, PhD. CreateSpace and Free Radical Publishing, © 2015).

(Howes, Pomegranate, broccoli, turmeric, 2015) Pomegranate, Broccoli (sulforaphane) & Turmeric (Curcumin): From the food cart to the medicine cart. Prof. R.M. Howes MD, PhD. CreateSpace and Free Radical Publishing, © 2015).

(Hsieh et al, 1998) (Hsieh, C. Y., Santell, R. C., Haslam, S. Z., Helferich, W. G., Estrogenic effects of genistein on the growth of estrogen receptor-positive human breast cancer (MCF-7) cells in vitro and in vivo. Cancer Res. 1998, 58, 3833–3838).

(Huang et al, 1999) (Huang, T.T., Carlson, E.J., Raineri, I., Gillespie, A.M., Kozy, H., Epstein, C.J. The use of transgenic and mutant mice to study oxygen free radical metabolism. Ann. NY Acad. Sci. 893, 95–112 (1999).

(Huang et al, 2000) (Huang P., Feng L., Oldham E. A., Keating M. J., Plunkett W. Superoxide dismutase as a target for the selective killing of cancer cells. Nature (Lond.), 407: 390-395, 2000).

(Hughes et al, 1998) (Hughes CM, Lewis SE, McKelvey-Martin VJ, Thompson W. The effects of antioxidant supplementation during Percoll preparation on human sperm DNA integrity. Hum Reprod. 1998 May;13(5):1240-7)

(Ikeno et al, 2005) (Ikeno Y, Hubbard GB, Lee S, et al. Housing density does not influence the longevity effect of calorie restriction. Journals of Gerontology Series a-Biological Sciences and Medical Sciences.2005;60:1510–7)

(Ishikawa et al, 2006) (Ishikawa et al. Smoking, alcohol drinking, green tea consumption and the risk of esophageal cancer in Japanese men. J Epidemiol. 2006 Sep;16(5):185-92).

(Islami et al, 2009) (Farhad Islami, Akram Pourshams, Dariush Nasrollahzadeh, Farin Kamangar, Saman Fahimi, Ramin Shakeri, Behnoush Abedi-Ardekani, Shahin Merat, Homayoon Vahedi, Shahryar Semnani, Christian C Abnet, Paul Brennan, Henrik Møller, Farrokh Saidi, Sanford M Dawsey, Reza Malekzadeh, Paolo Boffetta. Tea drinking habits and oesophageal cancer in a high-risk area in northern Iran: population-based case-control study. BMJ 2009; 338: b929).

(Ito et al, 1986) (Ito N, Hirose M, Fukushima S, et al. Studies on antioxidants: Their carcinogenic and modifying effects on chemical carcinogenesis. Food and Chemical Toxicology. Vol. 35, Issues 10-11, Oct-Nov 1986, pp 1071-1082).

(Jacobson, Raff, 1995) (Jacobson M. D., Raff M. C., Nature (London), 374, 814—816 (1995).

(Jenkins et al, 2008) (Jenkins JA, et al. Heterogeneity in randomized controlled trials of long chain (fish) omega-3 fatty acids in restenosis, secondary prevention and ventricular arrhythmias. J Am Coll Nutr. June 2008. Vol. 27. Nol 3. pp. 367-378)

(Johnson et al, 1996) (Johnson, T. M., Yu, Z. X., Ferrans, V. J., Lowenstein, R. A. & Finkel, T. (1996) Reactive oxygen species are downstream mediators of p53-dependent apoptosis. Proc. Natl. Acad. Sci. U.S.A. 93:11848-11852).

(Jorgensen, Gotzsche, 2006) (Jørgensen AW, Hilden J, Gøtzsche PC. Cochrane reviews compared with industry supported meta-analyses and other meta-analyses of the same drugs: systematic review. doi:10.1136/bmj.38973.444699.0B (Oct 6, 2006).

(Ju et al, 2008) (Ju, Y. H., Doerge, D. R., Woodling, K. A., Hartman, J. A. et al., Dietary genistein negates the inhibitory effect of letrozole on the growth of aromatase-expressing estrogen-dependent human breast cancer cells (MCF-7Ca) in vivo. Carcinogenesis 2008, 29, 2162–2168).

(Khan SI, et al, 2011) (Khan SI, et al, Potential utility of natural products as regulators of breast cancer-associated aromatase promoters. Reproductive Biology and Endocrinology. 2011,9:91)

(Kim Y.I., 2006) (Kim Y.I. Does a high folate intake increase the risk of breast cancer? Nutr. Rev. 2006;64:468–475).

(Kim et al, 2008) (Kim JE, Jin DH, Lee SD, Hong SW, Shin JS, Lee SK, Jun DJ, Kang JS, Lee WJ. Vitamin C inhibits p53-induced replicative senescence through suppression of ROS production and p38 MAPK activity. Int J Mol Med. 2008; 22:651–655).

(Kim et al, 2009) (Kim, D. J., Seok, S. H., Baek, M. W., Lee, H. Y. et al., Developmental toxicity and brain aromatase induction by high genistein concentrations in zebrafish embryos. Toxicol. Mech. Methods 2009, 19, 251–256).

(Kirk, 2007) (Kirk GD. Clinical Infectious Diseases, July 1, 2007).

(Klein EA, Thompson IM Jr, Tangen CM, et al., 2011) (Klein EA, Thompson IM Jr, Tangen CM, et al. Vitamin E and the risk of prostate cancer: The Selenium and Vitamin E Cancer Prevention Trial (SELECT). J Am Med Assoc. 2011; 306:1549–56).

(Klipstein-Grobusch K., Geleijnse J.M., den Breeijen J.H., et al., 1999) (Klipstein-Grobusch K., Geleijnse J.M., den Breeijen J.H., Boeing H., Hofman A., Grobbee D.E., Witteman J.C. Dietary antioxidants and risk of myocardial infarction in the elderly: The rotterdam study. Am. J. Clin. Nutr. 1999; 69:261–266).

(Knight, 1995) (Knight JA. Diseases related to oxygen-derived free radicals. Ann Clin Lab Sci. 1995 Mar-Apr;25(2):111-21).

(Koebel et al, 2007) (Koebel CM, Vermi W, Swann JB, Zerafa N, Rodig SJ, Old LJ, Smyth MJ, Schreiber RD. Adaptive immunity maintains occult cancer in an equilibrium state. Nature, 2007. Advance online publication (10.1038/Nature 06309).

(Konogaya et al.) (Konogaya M, et al. Clinical analysis of longstanding subacute myelo-optico-neuropathy: sequelae of clioquinol at 32 years after its ban. Journal of the Neurological Sciences. Volume 218, Issue 1, Pages 85-90, 15 March 2004).

(Kris-Etherton PM, Lichtenstein AH, Howard BV, et al., 2004) (Kris-Etherton PM, Lichtenstein AH, Howard BV, Steinberg D, Witztum JL. Antioxidant vitamin supplements and cardiovascular disease. Circulation. 2004;110:637–41).

(Krutchik et al, 1978) (Krutchik AN, Buzdar AU, Blumenschein GR, Lukeman JM. Spontaneous regression of breast carcinoma. Arch Intern Med. 1978;138(11):1734-1735).
(Kwak S, Myung S, Lee Y, Seo H., 2012) (Kwak S, Myung S, Lee Y, Seo H, Korean Meta-analysis Study Group, f Efficacy of omega-3 fatty acid supplements (eicosapentaenoic acid and docosahexaenoic acid) in the secondary prevention of cardiovascular disease: A meta-analysis of randomized, double-blind, placebo-controlled trials. Arch Intern Med. 2012;172:686–694).

(Kulling et al, 1999) (Kulling, S. E., Rosenberg, B., Jacobs, E., Metzler, M., The phytoestrogens coumoestrol and genistein induce structural chromosomal aberrations in cultured human peripheral blood lymphocytes. Arch. Toxicol. 1999, 73, 50–54)

(Kushi LH, Folsom AR, Prineas RJ, 1996) (Kushi LH, Folsom AR, Prineas RJ, Mink PJ, Wu Y, Bostick RM. Dietary antioxidant vitamins and death from coronary heart disease in postmenopausal women. N Engl J Med. 1996;334:1156–1162).

(Laranjinha J., 2009) (Laranjinha J. Oxidative Stress: From 1980's to Recent Update. In: Soares R., Costa C., editors. Oxidative Stress, Inflammation and Angiogenesis in the Metabolic Syndrome. Springer, Science + business Media; New York, NY, USA: 2009. pp. 21–32).

(Larsson et al, 2010) (Susanna C Larsson, Agneta Åkesson, Leif Bergkvist, and Alicja Wolk. Multivitamin use and breast cancer incidence in a prospective cohort of Swedish women. Am J Clin Nutr Published online 24 March 2010. Am J Clin Nutr Vol. 91, No. 5, 1268-1272, May 2010).

(Lavosier, 1781) (Lavosier AL. Considérations générales sur la nature des acides, et sur les principes dont ils sont composés Mémoires de l'Académie Royale des Sciences de Paris. 1781)

(Lawenda et al, 2007) (Lawenda BD, Smith DE, Xu L, et al. Do the dietary supplements epigallocatechin gallate or vitamin e cause a radiomodifying response on tumors in vivo? A pilot study with murine breast carcinoma. J Soc Integr Oncol (2007) 5(1):11–17).

(Lawenda et al, 2008) (Lawenda BD, Kelly KM, Ladas EJ, Sagar SM, Vickers A, Blumberg J. 2008. Should supplemental antioxidant administration be avoided during chemotherapy and radiation therapy? Journal of the National Cancer Institute. May 27, 2008. 100(11)773-783).

(Lawson et al, 2007) (Lawson KA, Wright ME, Subar A, Mouw T, Schatzkin A, Leitzmann MF. Multivitamin use and risk of prostate cancer in the National Institutes of Health–AARP Diet and Health Study. J Natl Cancer Inst (2007) 99:754–64).

(Ledesma et al, 2011) (Ledesma MC et al. Selenium and vitamin E for prostate cancer: post-SELECT (Selenium and Vitamin E Cancer Prevention Trial) status. Mol Med. 2011 Jan-Feb;17(1-2) :134-43)

(Lee IM, Cook NR, Manson JE, et al., 1999) (Lee IM, Cook NR, Manson JE, et al. Beta-carotene supplementation and incidence of cancer and cardiovascular disease: the Women's Health Study. J Natl Cancer Inst. 1999; 91:2102–6).

(Lee et al, 2004) (Duk-Hee Lee, Aaron R Folsom, Lisa Harnack, Barry Halliwell and David R Jacobs, Jr. Does supplemental vitamin C increase cardiovascular disease risk in women with diabetes? American Journal of Clinical Nutrition, Vol. 80, No. 5, 1194-1200, November 2004)

(Lee IM, Cook NR, Graziano JM, et al., 2005) (Lee IM, Cook NR, Graziano JM, et al. Vitamin E in the primary prevention of cardiovascular disease and cancer: The Women's Health Study: a randomized controlled trial. J Am Med Assoc. 2005; 294:56–65).

(Lees, 2003) (Lees KR, et al. Tolerability of NXY-059 at higher target concentrations in patients with acute stroke. Stroke 2003 Feb;34(2):482-7)

(Leppala et al. 2000) (Leppala JM, Virtamo J, Fogelholm R, Huttunen JK, Albanes D, Taylor PR, et al. Controlled trial of alpha-tocopherol and beta-carotene supplements on stroke incidence and mortality in male smokers. Arterioscler Thromb Vasc Biol 2000;20:230-5).

(Lesperance et al, 2002) (Lesperance ML, Olivotto IA, Forde N, et al. Mega-dose vitamins and minerals in the treatment of non-metastatic breast cancer: an historical cohort study. Breast Cancer Res Treat (2002) 76(2):137–143).

(Levy et al, 2004) (Levy AP, Friedenberg P, Lotan R, et al. The effect of vitamin therapy on the progression of coronary artery atherosclerosis varies by haptoglobin type in postmenopausal women. Diabetes Care. 2004;27(4):925-930)

(Li H, Stampfer MJ, Giovannucci EL, et al., 2004) (Li H, Stampfer MJ, Giovannucci EL, et al. A prospective study of plasma selenium levels and prostate cancer risk. J Natl Cancer Inst. 2004; 96:696–703).

(Lichtenberg, 2011) (Lichtenberg D. Who is likely to gain from high dose supplementation of vitamin E? Harefuah. 2011 Jan;150(1):37-40)

(Ling et al, 2003) (Ling Y-H, Liebes L, Zou Y and Perez-Soler R. Reactive Oxygen Species Generation and Mitochondrial Dysfunction in the Apoptotic Response to Bortezomib, a Novel Proteasome Inhibitor, in Human H460 Non-Small Cell Lung Cancer Cells. J. Biol. Chem., Vol. 278, Issue 36, 33714-33723, September 5, 2003).

(Liochev Sl, 2013) (Liochev Sl, Reactive oxygen species and the free radical theory of aging. Free Radic Biol Med. 2013 Jul; 60:1-4)

(Lippman SM, Klein EA, Goodman PJ, et al., 2009) (Lippman SM, Klein EA, Goodman PJ, et al. Effect of selenium and vitamin E on risk of prostate cancer and other cancers: The Selenium and Vitamin E Cancer Prevention Trial (SELECT). J Am Med Assoc. 2009; 301:39–51).

(Loeve et al, 2004) (Loeve F, Boer R, Zauber AG; et al. National Polyp Study data: evidence for regression of adenomas. Int J Cancer. 2004; 111(4):633-639).

(Long L.H., Hoi A., Halliwell B., 2010) (Long L.H., Hoi A., Halliwell B. Instability of, and generation of hydrogen peroxide by, phenolic compounds in cell culture media. Arch. Biochem. Biophys. 2010; 501:162–16).

(Lonn et al, 2005) (Effects of long-term vitamin E supplementation on cardiovascular events and cancer: a randomized controlled trial. E. Lonn et al. JAMA. 2005 Mar 16;293(11):1338-47)

(Losonczy K.G., Harris T.B., Havlik R.J., 1996) (Losonczy K.G., Harris T.B., Havlik R.J. Vitamin e and vitamin c supplement use and risk of all-cause and coronary heart disease mortality in older persons: The established populations for epidemiologic studies of the elderly. Am. J. Clin. Nutr. 1996;64:190–196).

(Lui et al, 2005) (Liu, B., Edgerton, S., Yang, X., Kim, A. et al., Low-dose dietary phytoestrogen abrogates tamoxifen-associated mammary tumor prevention. Cancer Res. 2005, 65, 879–886)

(Lundberg, 2015) (Lundbert GD. The certainty of uncertainty in medicine. Medscape Perspective May 14, 2015.)

(Ma Y, Chapman J, Levine M, et al., 2014) (Ma Y, Chapman J, Levine M, et al. High-dose parenteral ascorbate enhanced chemosensitivity of ovarian cancer and reduced toxicity of chemotherapy. Sci Transl Med. 2014; 6:222ra18).

(Magliano et al, 2006) (Magliano, Dianna; McNeil, John; Branley, Pauline; Shiel, Louise; Demos, Lisa; Wolfe, Rory; Kotsopoulos, Dimitra; McGrath, Barry. The Melbourne Atherosclerosis Vitamin E Trial (MAVET): a study of high dose vitamin E in smokers. European Journal of Cardiovascular Prevention & Rehabilitation. June 2006 - Volume 13 - Issue 3 - pp 341-347)

(Marwah, 2011) (Marwaj RK. Comorbidities in gouty arthritis. J Investig Med. 2011 Decd;59(8):1211-20)

(Maserejian et al, 2007) (Maserejian NW, Giovanncci E, Rosner B, Joshipura K. Prospective Study of Vitamins C, E, A, and Carotenoids and Risk of Oral Premalignant Lesions in Men. International J of Cancer. 120(5):970-7; 2006).

(Maurya D.K., Devasagayam T.P., 2010) (Maurya D.K., Devasagayam T.P. Antioxidant and prooxidant nature of hydroxycinnamic acid derivatives ferulic and caffeic acids. Food Chem. Toxicol. 2010;48:3369–3373).

(Maxwell, 1999) (Maxwell SR. Antioxidant vitamin supplements: update of their potential benefits and possible risks. Drug Saf. 1999 Oct; 21(4):253-66).

(Mayne et al, 2001) (Susan T. Mayne et al. Randomized Trial of Supplemental ß-Carotene to Prevent Second Head and Neck Cancer. Cancer Research 61, 1457-1463, February 15, 2001).

(McCord J.M., Keele B.B., Jr, Fridovich I., 1971) (McCord J.M., Keele B.B., Jr, Fridovich I. An enzyme-based theory of obligate anaerobiosis: The physiological function of superoxide dismutase. Proc. Natl. Acad. Sci. USA. 1971;68:1024–1027).

(McMichael-Phillips et al, 1998) (McMichael-Phillips, D. F., Harding, C., Morton, M., Roberts, S. A. et al., Effects of soy-protein supplementation on epithelial proliferation in the histologically normal human breast. Am. J Clin. Nutr. 1998, 68, 1431S–1435S).

(Med Sci Monit., 2002) (A radical approach to cancer. U. Das. Med Sci Monit. 2002 Apr;8(4):RA79-92).

(Meeker et al, 2010) (Meeker JD, et al. Semen quality and sperm DNA damage in relation to urinary bisphenol A among men from an infertility clinic. Reprod Toxicol. 2010 Dec;30(4):532-9).

(Miller et al., 2004) (Miller ER 3d, Pastor-Barriuso R, Dalal D, Riemersma RA, Appel LJ, Guallar E. Meta-analysis: high-dosage vitamin E supplementation may increase all-cause mortality. Ann Intern Med 2005;142:37-46).

(Miller et al, 2005) (Miller ER III, Pastor-Barriuso R, Dalal D, Riemersma RA, Appel LJ, Guallar E. Meta-analysis: high-dosage vitamin E supplementation may increase all-cause mortality. Ann Intern Med. 2005; 142:37-46).

(Miller et al, 2009) (Miller PE, Vasey JJ, Short PF et al. Dietary supplement use in adult cancer survivors. Oncol Nurs Forum. 2009 Jan;36(1):61-8).

(Miranda-Vilela et al, 2011) (Miranda-Vilela AL, et al. The protective effects of nutritional antioxidant therapy on Ehrlich solid tumor-bearing mice depend on the type of antioxidant therapy chosen: histology, genotoxicity and hematology evaluations. J Nutr Biochem. 2011 Jan 25. [Epub ahead of print]).

(Moch, 1986) (Moch RW. Pathology of BHA- and BHT-induced lesions. Food and Chemical Toxicology. Vol. 24, Issues 10-11, Oct-Nov 1986, pp 1167-1169).

(Moini H., Packer L., Saris N.E., 2002) (Moini H., Packer L., Saris N.E. Antioxidant and prooxidant activities of alpha-lipoic acid and dihydrolipoic acid. Toxicol. Appl. Pharmacol. 2002;182:84–90).

(Moscicki et al, 2004) (Moscicki AB, Shiboski S, Hills NK; et al. Regression of low-grade squamous intra-epithelial lesions in young women. Lancet. 2004;364(9446):1678-1683).

(Mow et al, 2002) (Mow BM, Chandra J, Svingen PA, et al. Effects of the Bcr/abl kinase inhibitors STI5 71 and adaphostin (NSC 680410) on chronic myelogenous leukemia cells in vitro. Blood. 2002;99: 664-671).

(Moyad et al. 2002) (Moyad MA. Selenium and vitamin E supplements for prostate cancer: evidence or embellishment? Urology. 2002 Apr;59(4 Suppl 1):9-19).

(MRC/BHF, 2002) (MRC/BHF Heart Protection Study of antioxidant vitamin supplementation in 20,536 high-risk individuals: a randomized placebo-controlled trial. Lancet. 2002 Jul 6;360(9326):23-33)

(Mursu J., Robien K., Harnack L.J., et al., 2011) (Mursu J., Robien K., Harnack L.J., Park K., Jacobs D.R., Jr Dietary supplements and mortality rate in older women: The iowa women's health study. Arch. Intern. Med. 2011;171:1625–1633).

(Myung, Int J Cancer. et al, 2009) (Myung et al, Green tea consumption and risk of stomach cancer: a meta-analysis of epidemiologic studies. Int J Cancer. 2009 Feb1;124(3):670-7).

(Myung S.K., Kim Y., Ju W., Choi H.J., 2010) (Myung S.K., Kim Y., Ju W., Choi H.J., Bae W.K. Effects of antioxidant supplements on cancer prevention: Meta-analysis of randomized controlled trials. Ann. Oncol. 2010;21:166–179).

(Myung S.K., Ju W., Kim S.C., 2011) (Myung S.K., Ju W., Kim S.C., Kim H. Vitamin or antioxidant intake (or serum level) and risk of cervical neoplasm: A meta-analysis. BJOG: Int. J. Obstet. Gynaecol. 2011;118:1285–1291).

(Nathan et al, vol 153, 1981) (Nathan CF et al. Tumor cell anti-oxidant defenses. Inhibition of the glutathione redox cycle enhances macrophage-mediated cytolysis. J Exp Med 1981 Apr 1;153(4): 766-82)

(Naughton et al. 2008) (Naughton F, Wijeysundera D, Karkouti K, Tait G, Beattie WS. N-acetylcysteine to reduce renal failure after cardiac surgery: a systematic review and meta-analysis. Can J Anaesth. 2008 Dec;55(12):827-35)

(Neil, Kay, 2006) (Neil E. Kay. ROS: double-edged sword for leukemic cells. (Mayo Clinic). Blood, 15 March 2006, Vol. 107, No. 6, pp. 2212-2213).

(NIH State-of-the Science Panel. 2007). (National Institutes of Health State-of-the-Science Conference Statement: Multivitamin/Mineral Supplements and Chronic Disease Prevention. NIH State-of-the-Science Panel. ANN INTERN MED 2006;145:364-371)

(Niki E, Noguchi N, Tsuchihashi H, Gotoh N., 1995) (Niki E, Noguchi N, Tsuchihashi H, Gotoh N. Interaction among vitamin C, vitamin E, and beta-carotene. Am J Clin Nutr. 1995; 62(6 Suppl):1322S–6S).

(Noto et al, 1989) (Noto V, Taper H S et al. Effects of sodium ascorbate (vitamin C) and 2-methyl-l, 4-naphthoquinone (vitamin K_3) treatment on human tumor cell growth in vitro. Cancer 1989; 63: 901 — 906).

(Omenn et al, 1996) (Omenn GS, Goodman GE, Thornquist MD, et al: Risk factors for lung cancer and for intervention effects in CARET, the Beta-Carotene and Retinol Efficacy Trial. J Natl Cancer Inst 88:1550-1559, 1996).

(Omenn GS, Goodman GE, Thorquist MD, et al., 1996) (Omenn GS, Goodman GE, Thorquist MD, et al. Effects of a combination of beta carotene and vitamin A on lung cancer and cardiovascular disease. N Engl J Med. 1996; 334:1150–5).

(Osganian SK, et al., 2003) (Osganian SK, et al. Vitamin C and risk of coronary heart disease in women. J Am Coll Cardiol. 2003;42:246–252. doi: 10.1016/S0735-1097(03)00575-8).

(Ozten et al, 2010) (Ozten N, Horton L, Lasano S, Bosland MC. Selenomethionine and alpha-tocopherol do not inhibit prostate carcinogenesis in the testosterone plus estradiol-treated NBL rat model. Cancer Prev Res (Phila). 2010 Mar;3(3):371-80).

(Padayatty et al, 2004) (Padayatty SJ, Sun H, Wang Y, et al. Vitamin C pharmacokinetics: implications for oral and intravenous use. Ann Intern Med (2004) 140(7):533–537).

(Padayatty et al, 2010) (Padayatty SJ, Sun AY, Chen Q, Espey MG, Drisko J, Levine M. Vitamin C: intravenous use by complementary and alternative medicine practitioners and adverse effects. PLoS One. 2010 Jul 7; 5(7): e11414).

(Page et al., 2010) (Page MM, Richardson J, Wiens BE, Tiedtke E, Peters CW, Faure PA, Burness G, Stuart JA: Antioxidant enzyme activities are not broadly correlated with longevity in 14 endotherm species. Age (Dordr). 2010, 32: 255-270).

(Parker-Pope, 2005) (Parker-Pope T. Cancer and Vitamins: Patients Urged to Avoid Supplements During Treatment; The Wall Street Journal 2005 Sep 20 Sect. D:1).

(Parrow NL, Leshin JA, Levin M., 2013) (Parrow NL, Leshin JA, Levin M. Parenteral ascorbate as a cancer therapeutic: a reassessment based on pharmacokinetics. Antioxid Redox Signal. 2013; 19:2141–56).

(Patterson RE, White E, Kristal AR, et al., 1997) (Patterson RE, White E, Kristal AR, et al. Vitamin supplements and cancer risk: the epidemiologic evidence. Cancer Causes and Control 1997; 8(5):786-802).

(Pechanova O., Simko F., 2009) (Pechanova O., Simko F. Chronic antioxidant therapy fails to ameliorate hypertension: Potential mechanisms behind. J. Hypertens Suppl. 2009;27: S32–S36) (Heart Protection Study Collaborative Group, 2002).

(Pelicano et al, 2003) (Inhibition of mitochondrial respiration: a novel strategy to enhance drug-induced apoptosis in human leukemia cells by a reactive oxygen species-mediated mechanism. H. Pelicano et al. J Biol Chem. 2003 Sep 26;278(39):37832-9).

(Peternelj T.T., Coombes J.S., 2011) (Peternelj T.T., Coombes J.S. Antioxidant supplementation during exercise training: Beneficial or detrimental? Sports Med. 2011;41:1043–1069).

(Piotrowska A., Bartnik E, 2014) (Piotrowska A., Bartnik E., The role of reactive oxygen species and mitochondria aging. Postepy Biochem. 2014;60(2):240-7).

(Pool-Zobel et al, 2000) (Pool-Zobel, B. L., Adlercreutz, H., Glei, M., Liegibel, U. M. et al., Isoflavonoids and lignans have different potentials to modulate oxidative genetic damage in human colon cells. Carcinogenesis 2000, 21, 1247–1252)

(Qiao YL, Dawsey SM, Kamangar F, et al., 2009) (Qiao YL, Dawsey SM, Kamangar F, et al. Total and cancer mortality after supplementation with vitamins and minerals: follow-up of the Linxian General Population Nutrition Intervention Trial [published erratum appears in J Natl Cancer Inst. 2010;102:140]. J Natl Cancer Inst. 2009; 101:507–18).

(Omenn G.S., Goodman G.E., Thornquist M.D., Balmes J., et al., 1996) (Omenn G.S., Goodman G.E., Thornquist M.D., Balmes J., Cullen M.R., Glass A., Keogh J.P., Meyskens F.L., Jr, Valanis B., Williams J.H., Jr, et al. Risk factors for lung cancer and for intervention effects in caret, the beta-carotene and retinol efficacy trial. J. Natl. Cancer Inst. 1996;88:1550–1559).

(Poljsak B., et al., 2005) (Poljsak B., Gazdag Z., Jenko-Brinovec S., Fujs S., Pesti M., Belagyi J., Plesnicar S., Raspor P. Pro-oxidative vs antioxidative properties of ascorbic acid in chromium(vi)-induced damage: An in vivo and in vitro approach. J. Appl. Toxicol. 2005;25:535–548).

(Printz 2001) (Printz C. Spontaneous regression of melanoma may offer insight into cancer immunology. J Natl Cancer Inst. 2001;93(14):1047-1048).

(Qiao et al, 2009) (Y.-L. Qiao, S. M. Dawsey, F. Kamangar, J.-H. Fan, C. C. Abnet, X.-D. Sun, L. L. Johnson, M. H. Gail, Z.-W. Dong, B. Yu, et al. Total and Cancer Mortality After Supplementation with Vitamins and Minerals: Follow-up of the Linxian General Population Nutrition Intervention Trial. J Natl Cancer Inst, April 1, 2009; 101(7): 507 - 518).

(Raal et al, 1999) (Efficacy of vitamin E compared with either simvastatin or atorvastatin in preventing the progression of atherosclerosis in homozygous familial hypercholesterolemia. Raal FJ, Pilcher GJ, Veller MG, Kotze MJ, Joffe BI. Am J Cardiol. 1999 Dec 1;84(11):1344-6, A7)

(Rahman I, Biswas SK, Kirkham PA., 2006) (Rahman I, Biswas SK, Kirkham PA. Regulation of inflammation and redox signaling by dietary polyphenols. Biochem Pharmacol. 2006 Nov 30;72(11):1439-52).

(Raitakari OT, et al., 2000) (Raitakari OT, et al. Oral vitamin C and endothelial function in smokers: Short-term improvement, but no sustained beneficial effect. J Am Coll Cardiol. 2000; 35: 1616–1621).

(Rapola et al, 1997) (Rapola, J.M., J. Virtamo, S. Ripatti, J.K. Huttunen, D. Albanes, P.R. Taylor & O. P. Heinonen. 1997. Randomized trial of alpha-tocopherol and beta-carotene supplements on incidence of major coronary events in men with previous myocardial infarction. Lancet. 349(9067):1715-1720)

(Rapola et al, 1998) (J M Rapola, J Virtamo, S Ripatti, J K Haukka, J K Huttunen, D Albanes, P R Taylor, and O P Heinonen. Effects of alpha tocopherol and beta carotene supplements on symptoms, progression, and prognosis of angina pectoris
Heart, May 1, 1998; 79(5): 454 - 458)

(Riatt et al, 2005) (Riatt et al, Fish oil supplementation and risk of ventricular tachycardia and ventricular fibrillation in patients with implantable defibrillators. JAMA. 2005;293(23):2884-2891)

(Rimm E.B., Stampfer M.J., Ascherio A., et al., 1993) (Rimm E.B., Stampfer M.J., Ascherio A., Giovannucci E., Colditz G.A., Willett W.C. Vitamin e consumption and the risk of coronary heart disease in men. N. Engl. J. Med. 1993;328:1450–1456).

(Riordan, et al, 1995) (N. H. RIORDAN, H. D. RIORDAN, X. MENG, Y. LI and J. A. JACKSON. Medical Hypotheses (1995), Volume 44, Number 3, March 1995, pp. 207-213).

(Ristow M., Zarse K., Oberbach A., et al., 2009) (Ristow M., Zarse K., Oberbach A., Kloting N., Birringer M., Kiehntopf M., Stumvoll M., Kahn C.R., Bluher M. Antioxidants prevent health-promoting effects of physical exercise in humans. Proc. Natl. Acad. Sci. USA. 2009;106:8665–8670).

(Ristow M., Schmeisser S., 2011) (Ristow M., Schmeisser S. Extending life span by increasing oxidative stress. Free Radic. Biol. Med. 2011;51:327–336).

(Rizos EC, Ntzani EE, Bika E, et al, 2012) (Rizos EC, Ntzani EE, Bika E, Kostapanos MS, Elisaf MS. Association between omega-3 fatty acid supplementation and risk of major cardiovascular disease events: A systematic review and meta-analysis. JAMA. 2012;308:1024–1033).

(Rodriguez et al, 2011) (Rodriguez KA, Wywial E, Perez VI, Lambert AJ, Edrey YH, Lewis KN, Grimes K, Lindsey ML, Brand MD, Buffenstein R. Walking the oxidative stress tightrope: a perspective from the naked mole-rat, the longest-living rodent. Curr Pharm Des. 2011;17(22):2290-307)

(Roswall et al, 2010, Lung cancer) (Roswall N, et al. Source-specific effects of micronutrients in lung cancer prevention. Lung Cancer. 2010 Mar;67(3):275-81).
(Rumbold et al, Apr 18, 2005. CD004072) (Rumbold A, Middleton P, Crowther CA. Vitamin supplementation for preventing miscarriage. Cochrane Database Syst Rev. 2005 Apr 18;(2):CD004069).

(Rumbold et al, Apr 18, 2005. CD004072) (Rumbold A, Crowther CA. Vitamin E supplementation in pregnancy. Cochrane Database Syst Rev. 2005 Apr 18;(2):CD004069).

(Rumbold et al, 2006) (Vitamins C and E and the Risks of Preeclampsia and Perinatal Complications Alice R. Rumbold, Ph.D., Caroline A. Crowther, Ross R. Haslam, Gustaaf A. Dekker, and Jeffrey S. Robinson, for the ACTS Study Group. N Engl J Med. 2006 Apr 27;354(17):1796-806)

(Rytter E., Vessby B., Asgard R., et al., 2010) (Rytter E., Vessby B., Asgard R., Ersson C., Moussavian S., Sjodin A., Abramsson-Zetterberg L., Moller L., Basu S. Supplementation with a combination of antioxidants does not affect glycaemic control, oxidative stress or inflammation in type 2 diabetes subjects. Free Radic. Res. 2010;44:1445–1453).

(Salonen JT, et al., 2000) (Salonen JT, et al. Antioxidant Supplementation in Atherosclerosis Prevention (ASAP) study: a randomized trial of the effect of vitamins E and C on 3-year progression of carotid atherosclerosis. J Intern Med. 2000;248:377–86).

(Sassi-Messai et al, 2009) (Sassi-Messai, S., Gibert, Y., Bernard, L., Nishio, S. et al., The phytoestrogen genistein affects zebrafish development through two different pathways. PloS ONE 2009, 4, e4935)

(Satia et al, 2009) (Satia JA, Littman A, Slatore CG, Galanko JA, White E. Long-term use of beta-carotene, retinol, lycopene, and lutein supplements and lung cancer risk: results from the VITamins and Lifestyle (VITAL) study. Am J Epidemiol. 2009 Apr 1;169(7):815-28).

(Sayin et al, 2014) (Sayin, V. I. et al. Antioxidants accelerate lung cancer progression in mice. Sci. Transl. Med. 6, 221ra15 (2014).

(Schilling et al, 2002) (Schilling FH, Spix C, Berthold F; et al. Neuroblastoma screening at one year of age. N Engl J Med. 2002;346(14):1047-1053).

(Schlecht et al, 2003) (Schlecht NF, Platt RW, Duarte-Franco E; et al. Human papillomavirus infection and time to progression and regression of cervical intraepithelial neoplasia. J Natl Cancer Inst. 2003;95(17):1336-1343).

(Schopel et al, 2013) (M. Schopel, Jockers, Autzen et al, (2013): Bisphenol A binds to Ras proteins and competes with Guanine Nucleotide exchange: implications for GTPase-selective antagonists, Journal of Medicinal Chemistry, 56(23):9664-72).

(Schurks et al. 2010) (Effects of vitamin E on stroke subtypes: meta-analysis of randomised controlled trials. Markus Schurks, Robert J Glynn, Pamela M. Rist, Christophe Tzourio, Tobias Kurth. BMJ 2010; 341:c5702 (online 11-4-10).

(Schwingshakl L, Hoffmann G., 2016) (Schwingshakl L, Hoffmann G. Does a Mediterranean-type diet reduce cancer risk? Curr Nutr Rep. 2016; 5:9–17).

(Semba et al., 2014) (Semba RD, Ferrucci L, Bartali B, Urpí-Sarda M, Zamora-Ros R, Sun K, Cherubini A, Bandinelli S, Andres-Lacueva C. Resveratrol Levels and All-Cause Mortality in Older Community-Dwelling Adults. JAMA Intern Med. 2014 May 12).

(Shafique et al, 2012) (Kashif Shafique, Philip McLoone, Khaver Qureshi, Hing Leung, Carole Hart, David S Morrison. Tea consumption and the risk of overall and grade specific prostate cancer: a large prospective cohort study of Scottish men. Nutr Cancer 2012 Aug 14;64(6):790-7. Epub 2012 Jun 14).

(Shargorodsky et al, 2010) (M. Shargorodsky, O. Debby, Z. Matas, R. Zimlichman. Effect of long-term treatment with antioxidants (vitamin C, vitamin E, coenzyme Q10 and selenium) on arterial compliance, humoral factors and inflammatory markers in patients with multiple cardiovascular risk factors. Nutr Metab, 7 (2010), p. 55).

(Shike et al, 2014) (Moshe Shike, Ashley S. Doane, Lianne Russo, Rafael Cabal, Jorge S. Reis-Filho, William Gerald, Hiram Cody, Raya Khanin, Jacqueline Bromberg and Larry Norton. JNCI J Natl Cancer Inst (2014) 106(9): dju189doi: 10.1093/jnci/dju189).

(Sigounas G, Anagnostu A, Steiner M., 1997) (Sigounas G, Anagnostu A, Steiner M. dl-Alpha tocopherol induces apoptosis in erythroleukemia, prostate, and breast cancer cells. Nutr Cancer. 1997; 28:30–5)

(Singh et al, 2005) (Singh SV, et al. Sulforaphane-induced Cell Death in Human Prostate Cancer Cells Is Initiated by Reactive Oxygen Species. J. Biol. Chem., Vol. 280, Issue 20, 19911-19924, May 20, 2005).

(Skrha et al. 1997) (J. Skrha, G. Sindelka and J. Hilgertova. The effect of fasting and vitamin E on insulin action in obese type 2 diabetes mellitus. Annals of the New York Academy of Sciences, Vol 827, Issue 1 556-560, 1997).

(Slater et al, 1995) (Slater, A. F., Nobel, C. S. & Orrenius, S. (1995) The role of intracellular oxidants in apoptosis. Biochim. Biophys. Acta 1271:59-62).

(Slatore et al, 2008) (Christopher G. Slatore, Alyson J. Littman, David H. Au, Jessie A. Satia, and Emily White Long-Term Use of Supplemental Multivitamins, Vitamin C, Vitamin E, and Folate Does Not Reduce the Risk of Lung Cancer. Am. J. Respir. Crit. Care Med. 2008 Mar. 1;177(5): 524-530).

(Soni et al. 2010) (Soni et al. Safety of vitamins and minerals: controversies and perspective. Toxicol Sci. 2010 Dec:118(2):348-55).

(Spallholz,1993) (Spallholz, JE, Zhang X, Boylan LM (1993) Generation of oxyradicals by selenium compounds. FASEB J. A290: 1683).

(Spallholz, 1994) (Spallholz, J.E. On the natue of selenium toxicity and carcinostatic activity. Free Radic. Biol Med. 1994. 17:45-64.).

(Spallholz, 2001) (Spallholz, J. E, Selenium and the prevention of cancer. Part II: Mechanisms for the carcinostatic activity of Se compounds. The bulletin of Selenium-Tellurium development association ISSN 1024-4204. October 2001.).

(Stampfer M.J., Hennekens C.H., Manson J.E., et al., 1003) (Stampfer M.J., Hennekens C.H., Manson J.E., Colditz G.A., Rosner B., Willett W.C. Vitamin e consumption and the risk of coronary disease in women. N. Engl. J. Med. 1993;328:1444–1449).

(Steinhubl, 2008) (Steven R Steinhubl. Why Have Antioxidants Failed in Clinical Trials? Article · Literature Review in The American Journal of Cardiology 101(10A):14D-19D · June 2008).

(Stevens et al, 2005) (Stevens VL, McCullough Ml, Diver WR, Rodriguez C, Jacobs EJ, Thun MJ, Calle EE. Use of multivitamins and prostate cancer mortality in a large cohort of US men. Cancer Causes Control. 2005 Aug; 16(6):643-50).

(Stolzenberg-Solomon R.Z., Sheffler-Collins S., Weinstein S., et al., 2009) (Stolzenberg-Solomon R.Z., Sheffler-Collins S., Weinstein S., Garabrant D.H., Mannisto S., Taylor P., Virtamo J., Albanes D. Vitamin e intake, alpha-tocopherol status, and pancreatic cancer in a cohort of male smokers. Am. J. Clin. Nutr. 2009;89:584–591).

(Stuart JA, et al., 2014) (Stuart JA, et al., A midlife crisis for the mitochondrial free radical theory of aging. Longevity & Healthspan20143:4).

(Sugiyama et al, 1996) (Sugiyama, H., Kashihara, N., Makino, H., Yamasaki, Y. & Ota, Z. (1996) Reactive oxygen species induce apoptosis in cultured human mesangial cells. J. Amer. Soc. Nephrol. 7:2357-2363).
(Suksomboon N., Poolsup N., Sinprasert S., 2011) (Suksomboon N., Poolsup N., Sinprasert S. Effects of vitamin e supplementation on glycaemic control in type 2 diabetes: Systematic review of randomized controlled trials. J. Clin. Pharm. Ther. 2011;36:53–63).

(Syu G.D., Chen H.I., Jen C.J., 2011) (Syu G.D., Chen H.I., Jen C.J. Severe exercise and exercise training exert opposite effects on human neutrophil apoptosis via altering the redox status. PLoS One. 2011;6:e24385).

(Tardif et al, 1997) (Tardif, J.C., Cote, G and Lesperance, J., et al. Probucol and multivitamins in the prevention of restenosis after coronary angioplasty. Multivitamins and Probucol Study Group. N Engl J Med 1997; 337(6): 365-372).

(Teicher et al, 1990) (Teicher, B.A, Holden, S.A., Al-Achi, A. and Herman, T.S. Classification of antineoplastic treatments by their differential toxicity toward putative oxygenated and hypoxic tumor subpopulations in vivo in the FSaII murine fibrosarcoma. Cancer Res 1990; 50: 3339-3344)

(The Alpha-Tocopherol Beta Carotene Cancer Prevention Study Group, 1994) (The Alpha-Tocopherol Beta Carotene Cancer Prevention Study Group. The effect of vitamin e and beta carotene on the incidence of lung cancer and other cancers in male smokers. The alpha-tocopherol, beta carotene cancer prevention study group. N. Engl. J. Med. 1994;330:1029–1035).

(Todd S., Woodward M., Tunstall-Pedoe H., et al., 1999) (Todd S., Woodward M., Tunstall-Pedoe H., Bolton-Smith C. Dietary antioxidant vitamins and fiber in the etiology of cardiovascular disease and all-causes mortality: Results from the scottish heart health study. Am. J. Epidemiol. 1999;150:1073–1080).

(Thompson, 1995) (Thompson, C. B. (1995) Apoptosis in the pathogenesis and treatment of disease. Science 267:1456-1462)

(Thörnwall et al., 2004) (Effect of α-tocopherol and ß-carotene supplementation on coronary heart disease during the 6-year post-trial follow-up in the ATBC study. Markareetta E. Törnwall et al. European Heart Journal 2004 25(13):1171-1178)

(Tonkonogi et al, 2003) (Reduced oxidative power but unchanged antioxidative capacity in skeletal muscle from aged humans. Tonkonogi, M. et al. Pflugers Arch. 2003 May;446(2):261-9. Epub 2003 Mar 27).

(Tworoger et al, 2011) (Tworoger et al, The combined influence of multiple sex and growth hormones on risk of postmenopausal breast cancer: a nested case-control study. Breast Cancer Res. 2011 Oct 21;13(5): R99).

(Valko et al, 2006) (Valko M, et al. Free radicals and antioxidants in normal physiological functions and human disease. Int J Biochem Cell Biol. 2007;39(1):44-84. Epub 2006 Aug 4)

(Valdiglesias et al, 2009) (Valdiglesias, V., Pasaro, E., Mendez, J., Laffon, B., In vitro evaluation of selenium genotoxic, cytotoxic, and protective effects: a review. Arch. Toxicol. 2009, 84, 337–351).

(Valko M., Leibfritz D., Moncol J., Cronin M.T., Mazur M., Telser J., 2007) (Valko M., Leibfritz D., Moncol J., Cronin M.T., Mazur M., Telser J. Free radicals and antioxidants in normal physiological functions and human disease. Int. J. Biochem. Cell Biol. 2007; 39:44–84).

(Vanhees et al, 2011) (Vanhees K, et al. Prenatal exposure to flavonoids: Implication for cancer risk. Toxicol Sci. 2011 Mar;120(1):59-67).

(Veal E.A., Day A.M., Morgan B.A., 2007) (Veal E.A., Day A.M., Morgan B.A. Hydrogen peroxide sensing and signaling. Mol. Cell. 2007;26:1–14).

(Vega-Lopez et al, 2004) (S. Vega-López, N. Kaul, S. Devaraj, R.Y. Cai, B. German, I. Jialal. Supplementation with ω3 polyunsaturated fatty acids and all-rac alpha-tocopherol alone and in combination failed to exert an anti-inflammatory effect in human volunteers Metabolism, 53 (2004), pp. 236-240).

(Virtamo J, Pientinen P, Huttunen JK, et al., 2003) (Virtamo J, Pientinen P, Huttunen JK, et al. Incidence of cancer and mortality following alpha-tocopherol and beta-carotene supplementation: a postintervention follow-up. J Am Med Assoc. 2003; 290:476–85).

(Villanueva and Kross, 2012) (Villanueva and Kross, Antioxidant-Induced Stress. Int J Mol Sci. 2012; 13(2): 2091–2109).

(Vivekanagthan et al, 2003) (Vivekananthan DP, Penn MS, Sapp SK, Hsu A, Topol EJ. Use of antioxidant vitamins for the prevention of cardiovascular disease: meta-analysis of randomised trials. Lancet. 2003; 361:2017-2023).

(Vrablic et al, 2001) (Vrablic, A. S., Albright, C. D., Craciunescu, C. N., Salganik, R. I. & Zeisel, S. H. (2001) Altered mitochondrial function and over generation of reactive oxygen species precede the induction of apoptosis by 1-O-octadecyl-2-methyl-rac-glycero-3-phosphocholine in p53-defective hepatocytes. FASEB J 15:1739-1744)

(Paul Emile Voest, Univ Med Ctr Utrecht, Cancer Cell 2011)

(Walsh et al., 2014) (Walsh ME, Shi Y, Van Remmen H: The effects of dietary restriction on oxidative stress in rodents. Free Radic Biol Med. 2014, 66: 88-99).

(Ward N.C., Hodgson J.M., Croft K.D., 2005) (Ward N.C., Hodgson J.M., Croft K.D., Burke V., Beilin L.J., Puddey I.B. The combination of vitamin c and grape-seed polyphenols increases blood pressure: A randomized, double-blind, placebo-controlled trial. J. Hypertens. 2005; 23:427–434).

(Ward, Wu et al, 2007) (Ward NC, Wu JH, Clarke MW, Puddey IB, Burke V, Croft KD, Hodgson JM. The effect of vitamin E on blood pressure in individuals with type 2 diabetes: a randomized, double-blind, placebo-controlled trial. J Hypertens. 2007 Jan;25(1):227-34)

(Waters et al, 2002) (Waters DD, Alderman EL, Hsia J, Howard BV, Cobb FR, Rogers WJ, Ouyang P, Thompson P, Tardif JC, Higginson L, Bittner V, Steffes M, Gordon DJ, Proschan M, Younes N, Verter JI. Effects of hormone replacement therapy and antioxidant vitamin supplements on coronary atherosclerosis in postmenopausal women: a randomized controlled trial. JAMA. 2002; 288: 2432–40)

(Watkins et al, 2000) (Watkins et al, Multivitamin Use and Mortality in a Large Prospective Study. American Journal of Epidemiology 152(2) · August 2000).

(Weinstein S.J., Wright M.E., Lawson K.A., et al. 2007) (Weinstein S.J., Wright M.E., Lawson K.A., Snyder K., Mannisto S., Taylor P.R., Virtamo J., Albanes D. Serum and dietary vitamin e in relation to prostate cancer risk. Cancer Epidemiol. Biomark. Prev. 2007; 16:1253–1259).

(Wentworth et al, 2001) (Wentworth P Jr, Jones LH, Wentworth AD, Zhu X, Larsen NA, Wilson IA, Xu X, Goddard WA 3rd, Janda KD, Eschenmoser A, Lerner RA. Antibody catalysis of the oxidation of water. Science. 2001 Sep 7;293(5536):1806-11)

(Wentworth et al, 2002) (Wentworth P Jr, McDunn JE, Wentworth AD, Takeuchi C, Nieva J, Jones T, Bautista C, Ruedi JM, Gutierrez A, Janda KD, Babior BM, Eschenmoser A, Lerner RA. Evidence for antibody-catalyzed ozone formation in bacterial killing and inflammation. Science. 2002 Dec 13;298(5601):2195-9).

(Williams et al, 2006) (F.E. Williams et al. Modulation by ellagic acid of DCA-induced developmental toxicity in the zebrafish (Danio rerio). J Biochem Mol Toxicol. 2006;20(4):183-90).

(Wray et al, 2009) (Wray DW, et al, Oral antioxidants and cardiovascular health in the exercise-trained and untrained elderly: a radically different outcome. Clin Sci (Lond). 2009 Mar;116(5):433-41)

(Wu et al, 2005) (Wu SJ, et al. Effects of antioxidants and caspase-3 inhibitor on the phenylethyl isothiocyanate-induced apoptotic signaling pathways in human PLC/PRF/5 cells. Eur J Pharmacol. 2005 Aug 22;518(2-3):96-106).

(Yamamoto et al, 1998) (Yamamoto K, Hanada R, Kikuchi A; et al. Spontaneous regression of localized neuroblastoma detected by mass screening. J Clin Oncol. 1998;16(4):1265-1269).

(Yokomizo et al, 1995) (Yokomizo A, Ono M, Nanri H, Makino Y, Ohga T, Wada M, Okamoto T, Yodoi J, Kuwano M, Kohno K. Cellular levels of thioredoxin associated with drug sensitivity to cisplatin, mitomycin C, doxorubicin, and etoposide. Cancer Res 1995;55:4293–4296)

(Yusuf S., Dagenais G., Pogue J., et al., 2000) (Yusuf S., Dagenais G., Pogue J., Bosch J., Sleight P. Vitamin e supplementation and cardiovascular events in high-risk patients. The heart outcomes prevention evaluation study investigators. N. Engl. J. Med. 2000;342:154–160).

(Zhang et al. 2009) (Zhang Y, Coogan P, Palmer JR, Strom BL, Rosenberg L. Vitamin and mineral use and risk of prostate cancer: the case-control surveillance study. Cancer Causes Control. 2009 Jul;20(5):691-8).

(Zhou et al, 2009 Cancer Res) (Zhuo, P., Goldberg, M., Herman, L., Lee, B. S. et al., Molecular consequences of genetic variations in the glutathione peroxidase 1 selenoenzyme. Cancer Res. 2009, 69, 8183–8190)